Submarine Channels

Processes & Architecture

Julian D. Clark & Kevin T. Pickering

Department of Geological Sciences,
University College London
Gower Street
London WC1E 6BT, UK

ISBN   0-9527313-0-4

We acknowledge Conoco U.K. Limited for their
financial support and in particular Grenville Timbrell
for his encouragement without which it is unlikely that
this book would have been published. The authors are
grateful to Charlie Bristow at Birkbeck College,
University of London, for reviewing the book which
resulted in considerable improvements to the text.
Assistance with the drafting of figures and contributions
in the field were gratefully received from Nick
Drinkwater, John Millington and Nigel Machin. Finally,
we would like to thank Mike Whitlock at Alden Press
for his unstinting help with the production of this book.

Published by
Vallis Press, London

Additional copies may be ordered from:
Vallis Press
Department of Geological Sciences
University College London
Gower Street
London
WC1E 6BT, UK

First published 1996

Printed by the Alden Press, Osney Mead, Oxford, UK

British Library Cataloguing in Publication Data
A catalogue record for this book is available from the
British Library

Cover: Photograph of stacked channel elements within
the Eocene Ainsa II submarine channel, south-central
Pyrenees, Spain.

Submarine Channels
Processes and Architecture

Julian D. Clark & Kevin T. Pickering

Contents

Part I: Introduction and architectural element analysis

Chapter 1: Introduction

Chapter 2: Submarine channel processes, architecture and depositional models

Chapter Three: Architectural elements

Part II: Modern systems

Chapter 4: Surveying and sampling seafloor sediments

Chapter 8: Quantitative analysis of ancient submarine channels

Part IV: Synthesis and channel models

Chapter 9: Synthesis

Chapter 10: Channel models

Chapter 11: Hydrocarbon prospectivity

Part I

Introduction

&

Architectural Element Analysis

Chapter 1

Introduction

Submarine channels have been documented from many ancient successions world-wide, the present-day seafloor and from areas of subsurface hydrocarbon exploration and development. With the exception of reservoir analogue studies, there have been few multidisciplinary studies of submarine channels, primarily due to difficulties in comparing datasets.

This book addresses the obvious gap in the scientific literature for a publication specifically on all aspects of submarine channels. This book is particularly timely because of the current academic and commercial interest in submarine channels, especially by those involved in hydrocarbon prospectivity.

An integrated approach to understanding submarine channels is adopted, from modern and ancient systems, and from a descriptive to quantitative perspective. In particular the range of channel architecture and channel-fill processes are presented to help evaluate the controls on such variations in a predictive manner, something that should find application in hydrocarbon exploration. Although the book focuses on submarine channels, data from other deep-marine sediment conduits such as canyons and gullies is also presented.

Channel morphology and sedimentology

Developments in the understanding of deep-marine channel geometry have progressed mainly from studies of modern fan systems. It is difficult to recognise true or complete channel planform and sectional area of ancient fan channels due to limited outcrop, post-depositional deformation and compaction, lack of good stratigraphic marker horizons, and the difficulty of distinguishing channel-fill muds from overbank muds.

Knowledge of deep-marine channels comes from both ancient successions and from modern channels. The sedimentology of ancient channel fills have been documented by many authors (e.g. Walker & Mutti 1973, Walker 1975a, 1978, Hein & Walker 1982, Pickering 1982a, 1985, Heller & Dickenson 1985, Nilsen 1985, Mutti *et al.* 1985, 1989, Mutti & Normark 1987, Pickering *et al.* 1986, 1995a). Shanmugam and Moiola (1985) distinguish five types of ancient channel fill based on the constituent facies (Figure 1.1).

There are many core and seismic facies descriptions of channels from modern submarine channels (e.g. Carlson & Nelson 1969, Griggs & Kulm 1970, Ness 1972, Nelson & Kulm 1973, Ness & Kulm 1973, Wilde *et al.* 1978, Piper & Normark 1983, Nelson 1985, O'Connell *et al.* 1985, Stow *et al.* 1985, Pickering *et al.* 1986a, Kolla &

Coumes 1987, Shanmugam *et al.* 1988) (Figure 1.2).

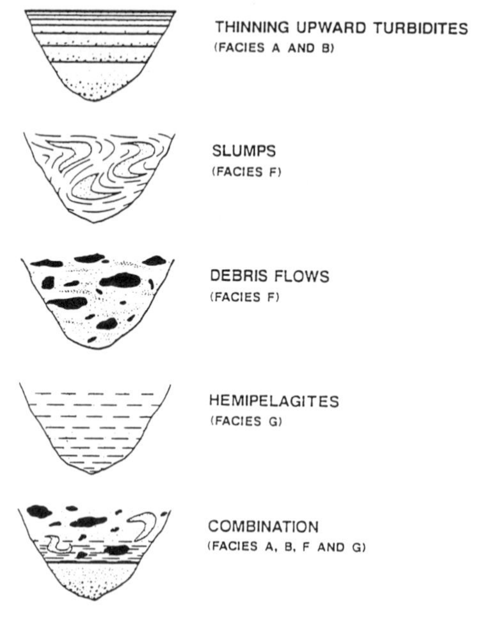

Figure 1.1. Submarine channel-fill models showing the variety of channel-fill facies (Shanmugam & Moiola 1985). Facies nomenclature refers to the facies scheme of Mutti and Ricci Lucchi (1972, 1975).

Modern deep-marine channel dimensions generally range from <10 km wide, and ca. 100 m deep (e.g. Upper Indus Fan channel, Kolla & Coumes 1987), to small distal-lobe channels <75m wide and with depths that are unresolvable on high frequency seismic profiles (<2 m), e.g. channels on the Outer Mississippi Fan (Twichell *et al.* 1991). Deep-marine channels occur on slope gradients between 1:133 and 1:4000 and may be "U" or "V" shaped (Nelson & Kulm 1973). Canyons may have greater widths and depths, and generally occur on slope gradients steeper than 1:200. The dimensions of some selected modern and ancient channel/canyon systems are shown in Table 1.1.

A levee is a morphological term to describe the positive relief features observed on some modern channel sections. Channel levees are formed by the overspill from channelised turbidity currents. On modern deep-marine channels, channel levees may be prominent or apparently absent and large-scale examples can easily be identified on high-resolution seismic profiles and show positive relief. Very little is known from ancient outcrops of channel levees as they are hard to distinguish from other "overbank" deposits mainly because they possess a high potential for differential compaction and post-depositional

Location of canyon/channel	Length (km)	Width (km)	Depth (m)	Levee width (km)	Reference
MODERN*					
Bengal Fan Bay of Bengal	Up to 3000	13-18	150-900	100	Curray and Moore (1974)
Indus Fan Arabian Sea	Up to 500	8-11	300-800	50	V. Kolla (1994, *pers comm*)
Amazon fan Equatorial Atlantic	Up to 250	3-15	250-600	50	Damuth and Flood (1983/1984)
Mississippi Fan Gulf of Mexico	Up to 400	2-15	150-450	50	Coleman *et al.* (1983), Garrison et al. (1982)
Rhone Fan Gulf of Lion	Up to 150	1-10	150-400	40	Droz (1983)
ANCIENT					
Doheny Channel Late Miocene, California	0.2	0.2	40+	-	Normark and Piper (1969)
Marnoso-Arenacea Fm. Middle Miocene, Italy	Up to 100	1.5	60-70	-	Ricci Lucchi (1981)
Hecho Basin Eocene, Northern Spain	3-4?	0.1	10	-	Mutti (1977)
Delaware Basin Permian, New Mexico/Texas	2-3?	0.4	15+	-	Jacka *et al.* (1968)
Shale Grit Fm. Late Carboniferous, England	3+	1	50	-	Walker (1966)

*Present-day deep-sea fans that were active during the Quaternary Period

Table 1.1. Modern and ancient fan canyon/channel dimensions. Redrawn after Shanmugam *et al.* (1985).

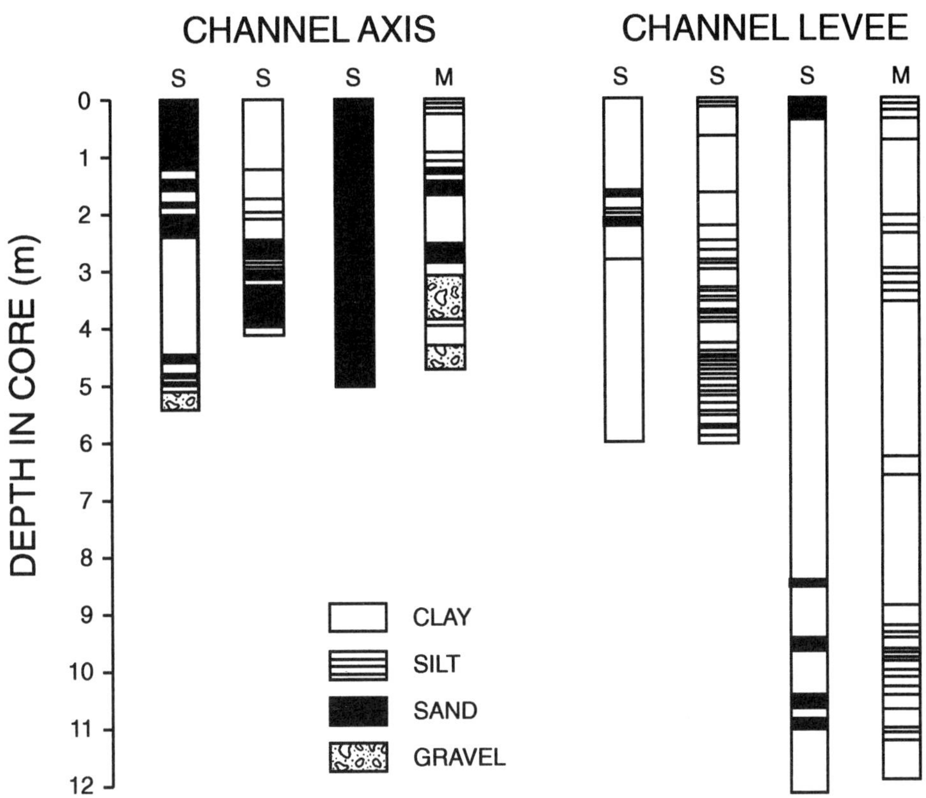

Figure 1.2. Piston cores from the channel axis and channel levee deposits of the Surveyor Channel (S) and Moresby Channel (M). Redrawn after Ness & Kulm (1973).

deformation (including faulting during burial and uplift). Because of the problem with recognising levees in ancient outcrops, possible levee deposits are grouped under the more general facies term "overbank" deposits, as described by Mutti and Normark (1987). Overbank facies commonly consist of fine-grained and thin-bedded, current-laminated sands and silts and also graded mudstones.

Levees are commonly higher on the right-hand-side looking down-channel in the Northern Hemisphere (opposite side for the Southern Hemisphere), due to the Coriolis Force acting on the channelised flow (Menard 1955, Komar 1969). Such levees may be up to 50 km wide and 300 m above the surrounding abyssal plain (e.g. Amazon Fan, Damuth *et al.* 1988).

Levee characteristics of upper and middle fan valleys, and non-fan channels, appear to have similar lithologies (Nelson & Kulm 1973). Levee deposits, as described by many authors (e.g. Normark *et al.* 1980, Damuth & Flood 1985, Kolla & Coumes 1987), typically show rhythmic alternations of laminated silty beds typically up to 3 cm thick, with some medium- to fine-grained sandy beds and lenses up to several centimetres.

In plan view, channels may be straight, sinuous, meandering, braided, anastomosed or more commonly show a combination of the above with down-channel changes in their planform geometry, resulting from changing seafloor gradient and topography. Tectonic features, such as faults and folds that directly affect the seafloor topography, commonly exert a strong influence of channel course and form. There is a relationship between the sinuosity and gradient of modern fan channels that permit the identification of high-sinuosity, low-gradient channel systems (e.g. Indus Fan channels), and low-sinuosity, high-gradient channel systems (e.g. Porcupine Seabight channels) (Clark *et al.* 1992).

In ancient channels, the observed facies characteristics have led researchers to describe channels as either braided, meandering, sinuous or straight (e.g. Hein & Walker 1982), but such a classification of planform geometry is generally not applicable to all ancient channel outcrops. There now exists, however, a sufficiently large published database of sonar data (both deep-towed and long-range) to accurately and quantitatively describe and classify many channel systems by their planform geometry - in a similar way to that used for fluvial channels. Lateral accretion and point-bar deposits have been documented from various meandering deep-marine channels (e.g. from the Mississippi channel, Pickering *et al.* 1986b). Channel bars produced by braided channels have also been documented (e.g. Watson 1981).

A classification of channel type is necessary to describe different channel fill-deposits in both ancient and modern submarine channels. The following sections outline four classification schemes for channel-type which have commonly been used in the literature to characterise channel morphology of modern and ancient submarine channels.

Erosional-depositional characteristics.

Modern and ancient channels can be characterised by the relative degrees of erosional and depositional architecture. The criteria for distinguishing depositional, erosional-depositional, and erosional channels is different between modern and ancient channels. For modern channels this classification is applied using the morphology of the (generally unfilled) channel-levee complex profile, whereas for ancient channels, the classification is based on the facies associations within the channel-fill and its relationship with adjacent levee deposits.

Channel position on the fan

Both modern and ancient deep-marine channels can be classified by their position on a fan. Additionally, some channels are classified as non-fan channels. In the fan models of Mutti and Ricci Lucchi (1972, 1975), and Normark (1970a, 1978), the upper fan regions of modern and inner fan areas of ancient fans can be considered as analogous. The channelised portion of the upper suprafan region of modern fans is comparable to the middle and upper channelised outer fan regions of ancient fans (Walker & Mutti 1973, Shanmugam & Moiola 1988) (Figure 1.3).

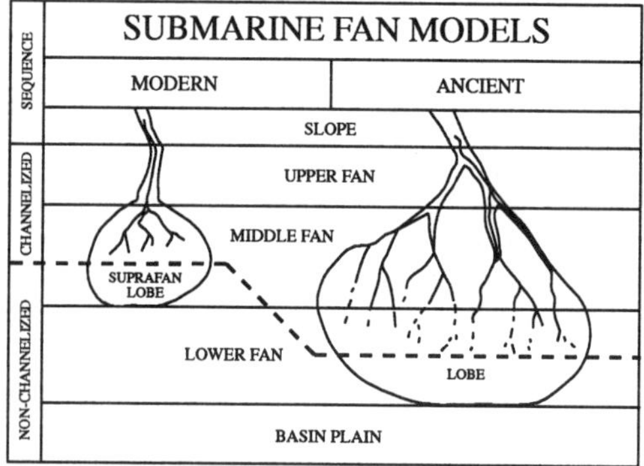

Figure 1.3. Comparison of terminology between ancient (Mutti & Ricci Lucchi 1972) and modern (Normark 1970a) fan models. Redrawn after Shanmugam & Moiola (1985).

Planform geometry

The planform geometry of modern submarine channel systems can be mapped with relative ease, and this permits a distinction between straight (sinuosity <1.1), sinuous (sinuosity =1.1-1.5), meandering (sinuosity >1.5), and braided channels. The sinuosity definitions given are those used in the classification of fluvial channels (see Schumm 1963). In ancient channel-levee complexes, it is seldom possible to make these distinctions from the mapping of individual channels without exceptionally good outcrop exposures, or tight well correlation in subsurface geology.

Channel-type related to relative changes in sea-level

Submarine channels typically have different morphology and other sedimentary characteristics, depending on the height of relative sea-level (with respect to the shelf/slope break) at the time of incision and, or, fill of the channel. Changes in relative base level (e.g. sea-level) commonly can be identified in the morphology and sedimentology of ancient deep-marine successions. In most modern deep-marine channels, the morphology and sedimentology can be shown to have been affected by the Holocene global (eustatic) rise in sea-level, and hence this classification is generally restricted to the classification of ancient submarine channels.

Table 1.2 summarises the four end-member classifications of modern and ancient channels. It is important to remember that different channel morphological types reflect various channel-flow processes, and since sedimentary processes within individual channels are likely to undergo down-channel changes, any channel-type classification may not be a unique classification for the entirety of the channel's length.

Channel type	Modern/Ancient
Depositional Depositional-erosional Erosional	modern & ancient
Slope Upper fan (inner fan) Middle fan Lower fan Non-fan	modern & ancient
Straight Sinuous Meandering Braided Anastomosed	generally for modern channels only
Lowstand Transitional Highstand	generally for ancient channels only

Table 1.2. Classification of modern and ancient deep-marine channel type.

Erosional-, depositional- and mixed-channel characteristics

Three basic types of channel cross-section are recognised based on the morphology of the entire channel-levee complex as seen on high-frequency seismic profiles (Hamilton 1967, Laughton 1968, Normark 1970a, Embley *et al.* 1970) (Figure 1.4).

- Depositional channels.
- Erosional-depositional (or modified) channels.
- Erosional channels.

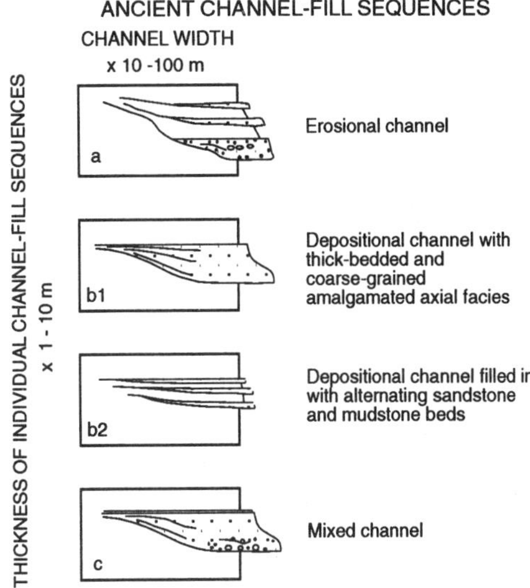

Figure 1.4. Ancient channel-fill models for erosional, depositional and mixed channel-fill types (Mutti & Normark 1987).

In some circumstances it is possible for deep-marine channels to undergo down-channel changes with respect to their erosional or depositional characteristics. The channel-canyon system on the north-west slope of the Gulf of Mexico, described by Satterfield and Behrens (1990) is an example of a modern channel which changes between a leveed aggradational channel form and an erosional channel form as it passes between intraslope basins separated by topographic highs formed by salt diapirism. Channels within the slope basins show depositional characteristics, but are connected by interbasin erosional channels. For ancient channels, Mutti and

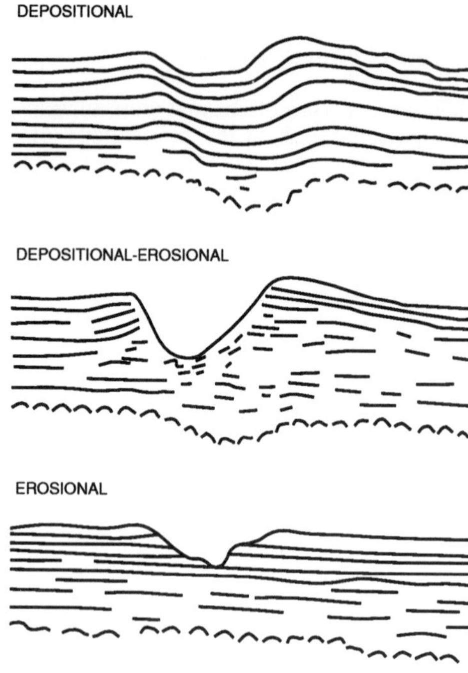

Figure 1.5. Profiles of depositional, depositional-erosional, and erosional modern submarine channels (Normark 1970a).

Normark (1987), have also identified erosional, depositional, and mixed channel-fill sequences (Figure 1.5). Care must be exercised in using such terms interchangeably, between modern and ancient environments, as the three ancient channel types are identified only from their channel-fill facies and architecture. Also, a channel may develop discrete erosional and depositional phases throughout the development of its growth and fill-history, and therefore could show repeated phases of erosion and deposition typical of the mixed-type channel. Hence, the ancient mixed-type channel may be a product of changes in the state of an individual channel's erosional-depositional character.

Modern depositional channels (or aggradational channels) can be recognised by the following features, seen on high-frequency seismic profiles (e.g. Hamilton 1967):

- Apparently continuous reflectors.
- The channel floor is elevated above the level of the surrounding fan/plain, and consists of flat lying reflectors.
- Levees are always present on both sides, the right-hand-side levee (looking down-channel in the Northern Hemisphere) is commonly higher and wider.
- Sub-bottom reflectors may suggest a shift in channel axis with time to the left hand side (looking down channel in the Northern Hemisphere): lateral migration to this side is favoured by the lower height of its levees.

Modern depositional channels are the result of deposition from various sediment gravity flows taking place in the channel axis simultaneously with the construction of levees resulting from overspill processes. Channel relief is the product of the balance between the amount of deposition in the channel axis and on the levees. It can, therefore, be assumed that channel relief is controlled by the hydrodynamics of the channelised flows (Nelson & Kulm 1973). Depositional channels have been documented from the middle and upper fan regions (e.g. Amazon Fan, Flood & Damuth 1987 (Figure 1.6), the aggradational upper Indus Fan channels, Kolla & Coumes 1987), and the lower fan (e.g. Hamilton 1967). These channel types are commonly associated with major river-fed submarine fans on oceanic crust.

Depositional channel facies can also be recognised in ancient channel deposits (Mutti 1977, Pickering 1982a, Ricci Lucchi 1969, Walker 1966, 1975b, Mutti & Normark 1987). These facies result from the depositional phase of turbidity currents where much of the sediment load is deposited in the channel rather than being carried basinward. The facies of these channel deposits are characterised as follows:

- Broad and shallow channel features.
- Thick extensively scoured and amalgamated graded sandstone and pebbly sandstone beds deposited in the channel axis.
- Thinner beds, and finer-grained deposits separated by increasingly thicker mudstone partings towards the channel margins.

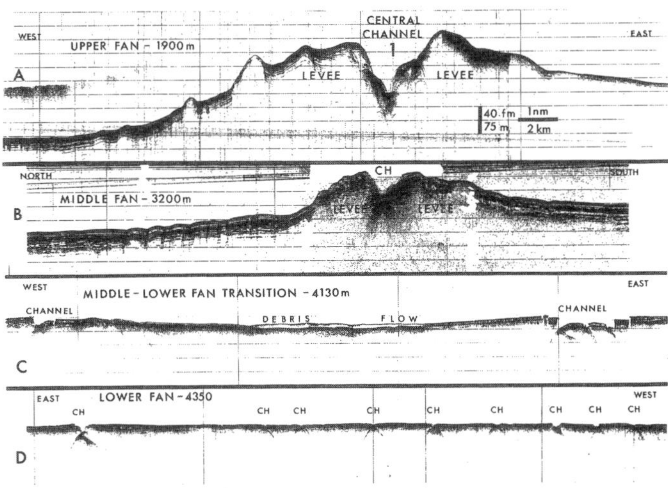

Figure 1.6. 3.5 kHz profiles of the Amazon Fan channel complexes. (a) Large, leveed upper fan channel, (b) leveed middle fan channel, and (c) unleveed lower fan channels (Flood & Damuth 1987).

The large-scale levees that characterise some modern depositional channels (e.g. The Upper Indus Channel) have not been recognised in ancient successions (Shanmugam *et al.* 1985).

Modern erosional channels are characterised by their morphology as shown on high-frequency seismic profiles. Nelson and Kulm (1973) described these characteristics as follows:

- Truncated horizontal beds in channel walls. These beds show no evidence for any constructional features adjacent to the channel.
- Terraces or benches formed by erosive downcutting within the channel axis.

These channels may be cut by erosive turbidity currents under specific hydrodynamic conditions, i.e. by flows that were not in hydrodynamic equilibrium with the seafloor due to constrictions imposed by seafloor topography (e.g. the Cascadia channel in the Blanco Fracture Zone (Griggs & Kulm 1970, Embley 1985), or from relatively recent changes in channel course (e.g. the Monterey channel, Normark 1970b: La Jolla Channel, Normark & Piper 1969, and the Rhône Fan channel, O'Connell *et al.* 1991).

In Mutti and Normark's (1987) channel facies description, ancient erosional channels are associated with coarse-grained deposits of the upper fan and slope areas, and these erosive channels may be characterised by the following:

- Coarse-grained and scoured facies.
- Lenticular bodies of clast supported conglomerates.
- Mud-supported conglomerates (associated with debris flows).
- Stratified and cross-stratified alternations of coarse-sandstone and granule-to-pebble conglomerates.
- Little or no overbank deposits associated with the above facies (a to d).
- Bank collapse sediment slides.

Examples of these channels are given by Carter and Norris (1977), Clifton (1984), Hein and Walker (1982), and Mutti *et al.* (1985, 1989).

For both modern and ancient systems, mixed erosional-depositional channels are the most common channel type , and they display some of the characteristics of both erosional and depositional channel types. Modern mixed-type channels are believed to be formed firstly under depositional conditions which later transform to erosive conditions, to result in the downcutting through older deposits (Heezen *et al.* 1969, Normark 1970a, Embley *et al.* 1970, Griggs & Kulm 1973). These channels are characterised by:

- Levees present on both sides, again one levee may be higher than the other (the right-hand-side levee looking down channel in the Northern Hemisphere).
- The asymmetry of the levees caused by the Coriolis force as revealed by sub-bottom reflectors, may suggest a shift in channel axis with time to the left-hand side (looking down channel in the Northern Hemisphere).
- Truncated beds may appear in both walls and levees.
- The channel floor may be higher or lower than the level of the surrounding seafloor.

In contrast to modern deep-marine channels, Mutti and Normark (1987) suggest that ancient mixed-type channels commonly show facies diagnostic of erosional conditions (expressed by coarse-grained residual deposits), followed by characteristically depositional facies. This cycle may be repeated within individual channel-fill complexes. Facies descriptions from outcrops of this channel type are given in Carter and Lindquist (1975), Dupuy *et al.* (1963), Mutti (1979), Walker (1985), and Cazzola *et al.* (1981), .

Mutti and Normark (1987) propose that after deposition within the above three channel types, the channels deposits are either abruptly overlain by mudstones, or grade into thin-bedded turbidites and fine-grained deposits with small-scale sandstone channel features. The former represents sudden channel abandonment, while the latter more common case represents the final and gradual abandonment of the channel and a decrease in flow volume. The small-scale channels may show localised lateral accretion (Mutti *et al.* 1985).

Growth and development

Deep-marine channel sedimentation has commonly been related to sequence stratigraphic concepts. In Mitchum's (1985) seismic stratigraphic paradigm the upper fan valley is a region of sediment bypass and

erosion during a lowstand. For this scenario, the channel would be described as "erosive". The channel levee complexes are subsequently formed during the initial rise of sea-level, and with a further rise in sea-level, successive back-filling of the channel with fine-grained sediments occurs (i.e. retrogressive displacement of the loci of major deposition up the channel). Note that different portions of a single channel may be in both a state of deposition and erosion simultaneously. For example, the upper fan valley of the Indus Fan is presently in a depositional state due to sea-level rise, while the shelf-slope feeder canyon is in a zone of degradation and erosion (Kolla & Macurda 1988).

Comparing ancient and modern: existing problems and application to integrated channel fill models

Careful consideration of the characteristics of basin type, size of sediment source, physical and temporal scales and stage of development, is important when comparing any deep-sea sedimentary systems (Mutti & Normark 1987). A further problem arises when making analogies between ancient and modern systems resulting from the very different methods of observation and data acquisition. Ensuring similar features are being compared in integrated studies is best achieved by looking at individual fan elements that can be recognised in both ancient and modern settings, e.g. channels, over-bank deposits, lobes, channel-lobe transition scours (Mutti & Normark 1987). It must be remembered, however, that the relationships between turbidite facies associations which characterise the ancient fan elements, and their related fan sub-environments, have rarely been confirmed from modern fans.

Type of basin and sediment source

Sediment source area and basin size exert a primary control on the development of channels on fans (Pickering 1982b, Kenyon 1992, Wetzel 1993). The largest modern fans (e.g. Indus, Amazon, Mississippi) are associated with large rivers and large basins. These fan systems are long-lived, and build up on passive continental margins. Channels are typically meandering, sinuous, and the sites of sand accumulation.

Ancient analogues of such large-scale muddy fan systems are only found in areas where accretionary prism tectonics have led to the uplift of the sediments from the seafloor, and consequently these sediments are highly deformed. In oceanic-tectonic settings, channel course will be tectonically controlled, e.g. Cascadia Channel (Griggs & Kulm 1970, Embley 1985), and the Surveyor Channel (Ness & Kulm 1973). In small active continental basins, such as the Eocene Hecho basin, Spanish Pyrenees, or modern Californian borderland basins, fan deposition is constrained and controlled by the development of its constituent channels (Mutti & Normark 1987). Kenyon (1992) suggested, that for a small or starved source area, fan channels have a low sinuosity and transport the sandy sediment to the channel-mouth lobes.

Stage of development

Observations from modern deep-marine channels largely reflect the state of channel development and sedimentation for one given time period, i.e. the present. Ancient and sub-surface channels are recognised by facies associations that provide a detailed account of the vertical sequence of sedimentation, which may represent complex stages of erosion and deposition, possibly repeated cycles.

Detailed investigations of the recent depositional history of modern fans, show remarkably different characteristics to those that are seen from present-day morphologies (e.g. the present-day morphology of the Mississippi Fan shows it to have a single point source, while subsurface seismic studies suggest that the Mississippi Fan was fed throughout the Quaternary Period by a multiple sediment source (Weimer & Buffler (1988) (Figure 1.7).

Scale and observation

The main problem arising from comparing ancient and modern channels is the lack of common datasets. For modern deep-sea sedimentary systems, most of the data comes from long-range side-scan sonar (e.g. GLORIA), with a maximum resolution of 50-100 m, narrow-beam and deep-towed sounding systems (e.g. TOBI, SeaBeam), maximum resolution 10-50 m, and seismic reflection profiles. High-energy, deep-penetration seismic reflection profiles are used to observe basin-scale features, with a minimum resolution of 30-50 m. High resolution, high frequency seismic profiles (e.g. 3.5 kHz) are used for more detailed sea-bed observations, but provide little penetration of the sea-bed and thick sands (>50 cm) may not be penetrated at all.

The internal geometry of fan elements such as channels (<30-50 m deep) may not be resolved on seismic reflection profiles (Mutti & Normark 1987). High-amplitude discontinuous reflectors underlying many modern channel floors are generally attributed to coarse-grained channel-base lag deposits (Bouma *et al.* 1985a, Normark & Gutmacher 1985, Stelting *et al.* 1985). Core samples from modern fans are generally inadequate for determining the sedimentology of deep-marine channels, since sandstone beds thicker than a few metres, and sediments coarser than gravels are rarely continuously sampled.

Problems associated with channels in ancient settings arise from:

- Lack of precise correlation to enable accurate "time slice" comparisons to be made with their modern counterparts.
- Effects of shale compaction adjacent to sandy deposits, i.e. differential compaction.
- Structural deformation.
- Limits of geological exposure for large-scale features, such as deep-marine channels.
- One-dimensional views in cores taken from subsurface reservoirs.

The largest continuous exposures are generally no more than 1 or 2 km in width. In contrast, some modern fan channel-levee systems are large enough to contain the outcrops of entire ancient turbidite systems, and at these scales such channel-levee systems have not been recognised from outcrop.

Ancient deep-marine channels can be detected on the highest scale of resolution of seismic data from subsurface fan systems. Channels with widths and depths less than this scale of resolution (generally 30-50 m) will not be detected. Well logs and core analysis provide a more detailed view of the sedimentology,

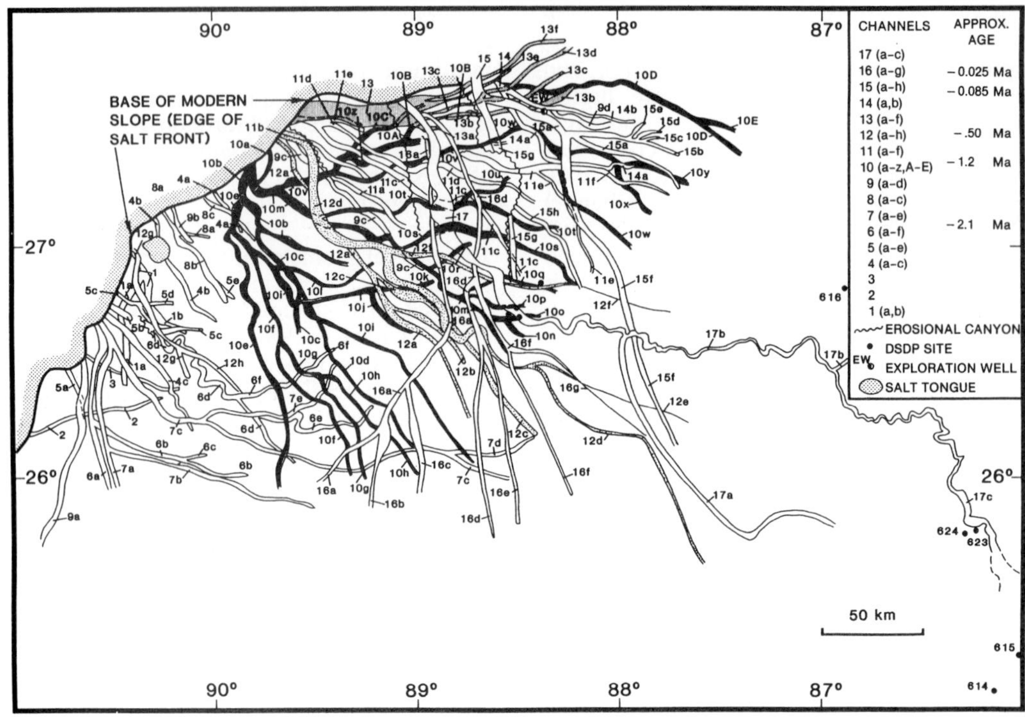

Figure 1.7. Map of 17 channel-levee systems (numbered 1-17) of the Plio-Pleistocene Mississippi Fan sediments, identified from multifold seismic data, and the youngest channel (17) mapped from sidescan sonar data. The map shows the variety of channel morphologies that developed since the Pliocene (after Weimer & Buffler 1988).

but lack information on lateral continuity or bedding geometry.

Chapter 2

Submarine channel processes, architecture and depositional models

Fluvial and submarine channel processes

The research results of geologists, geomorphologists and engineers studying the facies architecture, geomorphology and hydrodynamics of fluvial environments can aid in understanding submarine channel systems. This chapter reviews some basic processes acting in channelised flow, and the processes and facies of modern and ancient fluvial channels relevant for understanding of submarine channel processes. Similarities and differences between fluvial and submarine channels are highlighted.

From a knowledge of fluvial and submarine channel processes, and the ways in which these processes influence fluvial system geometry and depositional architecture, channel models for modern and ancient submarine channels are proposed. The sequence stratigraphic implications of channel-fill processes is also discussed with reference to such channel models (see also Chapter 10).

Channelised flow

There are two principal external forces acting on channelised flow. Gravity propels water downslope, and friction between channel boundaries resists the downslope movement. The frictional force is a shear stress proportional to the square of the flow velocity:

$$\text{Resistance } (k) = \tau / v^2 \qquad (2.1)$$

where τ is the shear stress per unit area and v is velocity. Shear stress varies throughout the height of a flow, from a maximum at the flow base to a minimum at the top of the flow:

$$\tau = K \, dv / dy \qquad (2.2)$$

where v is the velocity, y is the depth, or flow thickness and K is the molecular viscosity in viscous flow, and the eddy viscosity in turbulent flow. The flow velocity at the channel base is zero and increases upwards at a rate dv / dy. In laminar flow the mixing between laminar layers is molecular (i.e. by viscous forces), and in turbulent flow the mixing is by eddies.

The laminar or turbulent character of a flow is expressed in terms of the dimensionless Reynolds number (R) such that:

$$R = \frac{vd}{(\mu / \rho)} \qquad (2.3)$$

where d is the depth or flow thickness, μ is the dynamic viscosity, and ρ is the density.

The mean velocity of a channelised flow is a more useful parameter to measure experimentally. Discharge is the rate of flow passing through a section of the channel per unit time. Mean velocity is dependent on the energy gradient (slope), depth of flow, and channel boundary roughness.

The shear stress also can be related to the specific weight of the fluid (γ), the slope ($s = \sin \alpha$) and the hydraulic radius ($A / (2d+w)$), where A is the cross sectional area, and w is the channel width:

$$\tau = \gamma Rs \qquad (2.4)$$

By substituting equation (2.1) for shear stress into equation (2.4) it follows that:

$$v^2 = k\gamma Rs \qquad (2.5)$$

and if $\sqrt{(k\gamma)}$ is the Chezy coefficient of internal friction (C) then:

$$v = C\sqrt{(Rs)} \qquad (2.6)$$

In wide shallow channels, where d' is approximately the mean depth, then the mean channel velocity is proportional to the square root of the depth slope-product. Engineers commonly use an alternative version of the above equation, known as the Manning equation:

$$v = (1.49 \, d'^{2/3} \, s^{1/2}) / n \qquad (2.7)$$

Experimental values of n range from 0.01 for smooth metal surfaces, to 0.06 for natural channels with rough irregular channel beds.

The body velocity of a turbidity current (v_b) is also given by a Chezy-type equation:

$$v_b^2 = \frac{8g \, d_b \tan \alpha}{[f_o+f_i)] \, [\delta\rho / (\rho+\delta\rho)]} \qquad (2.8)$$

(Middleton 1966b)

where $\delta\rho$ = density difference between the density flow and seawater, ρ = density of seawater, d_b = body thickness, α = bottom slope in degrees, f_o = dimensionless Darcy-Weisbach friction coefficient for bed friction

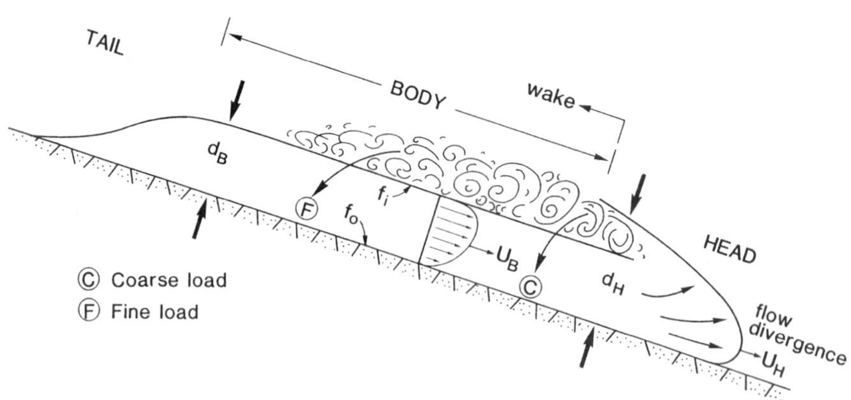

Figure 2.1. Head, body and tail regions of a large turbidity current (From Pickering *et al.* 1989).

and f_i = dimensionless friction coefficient for the interfacial friction at the top of the flow (Figure 2.1). The friction coefficients differ from those used for river and flume calculations, because there is a component of friction along the interface between the flow and the overlying water (f_i). For large turbidity currents (f_o+f_i) is approximately 0.01 (Middleton & Southard 1984).

Head velocity is equivalent to the body velocity on gentle slopes ($< 1\text{-}2°$), but on steeper slopes the ratio of the head velocity to body velocity < 1.0. In the latter case, the rate of body flow into the head region is balanced by the loss of suspended material in the region of intense turbulence, immediately behind the head (Simpson 1972).

Channelised turbidity currents

The three main differences between fluvial and deep-marine channelised flow are:

- Maximum flow depths in rivers only exceed bankfull depths during flood conditions, whereas it is common for turbidity currents to exceed the height of the channel levees.
- In rivers there is no entrainment of the overlying ambient fluid (air) into the flow.
- Many fluvial channels experience continuous-flow conditions, but submarine channels are more analogous to flash-flood or ephemeral stream channels.
- In submarine channels, it seems reasonable to predict that flow-stripping (*sensu* Piper & Normark 1983) at channel bends is relatively common unlike in fluvial environments.

Turbidity current initiation and flow

Turbidity currents can be initiated from the following processes: direct river discharge (hyperpycnal flow), storm surges, sediment failure (including seismically triggered failure), interception of sediment carried by longshore drift currents at gullies and canyon heads, and rip currents in canyons. Many natural turbidity currents are believed to have prolonged flow conditions i.e. quasi-steady flows, for many tens of hours (Normark &

Piper 1991). Prolonged flow may explain why regular sinuous channel morphology can develop.

Turbidity currents can continue to transport sediment over large distances in suspension, without deposition, which is achieved, theoretically, by a process known as autosuspension (Bagnold 1962). Once a flow is generated, a density contrast is set up between the ambient fluid, seawater, and the current containing entrained sediment, which results in downslope movement. The turbulence generated by a flow moving downslope maintains grains in suspension, which sustains the density contrast. True autosuspension may be rare in nature and is probably only achieved for sediment finer than fine sand and in thick flows above a critical density (Pantin 1979).

Flow characteristics in submarine channels

Flow characteristics and flow conditions can be deduced from channel morphology, such as channel dimensions, degree of channel wall erosion, lateral migration of channels and the morphology of large-scale depositional bedforms and erosional features within channels (Normark & Piper 1991). Menard (1964) showed how channel cross-section dimensions can be used to interpret flow conditions, principally by the down-slope changes in the flow thickness required, to allow a flow to maintain continuity of discharge (Figure 2.2a). Komar (1973) showed how the profile of down-channel relief may evolve towards an equilibrium (Figure 2.2b).

Normark and Piper (1991) attribute the wide range of channel characteristics observed in modern environments to the concentration and size of the turbidity currents and the type of sediment load (i.e. grain-size, density, etcetera). Muddy turbidity currents are generally thicker than sandy turbidity currents and therefore are more prone to channel overspilling (Normark & Piper 1991). The 1929 Grand Banks turbidity current for example was a predominantly sandy turbidity current, only a few 100 metres thick, and with a velocity >15 m s^{-1}, compared to muddy Pleistocene turbidity currents on the Laurentian Fan which appear to have spilled

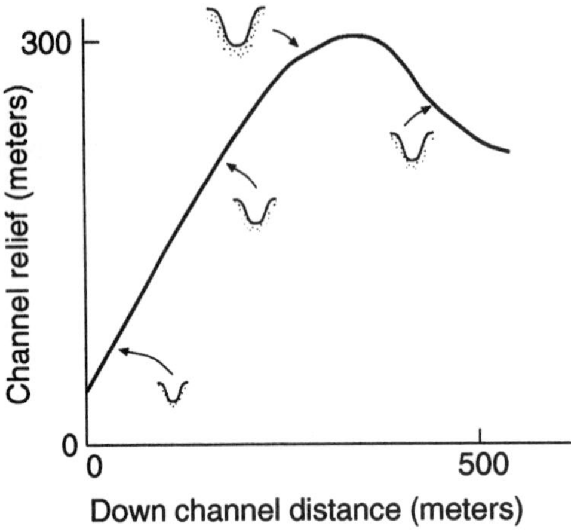

Down channel distance (meters)

Figure 2.2 (a). Down-channel variations in channel relief. Redrawn after Menard (1964).

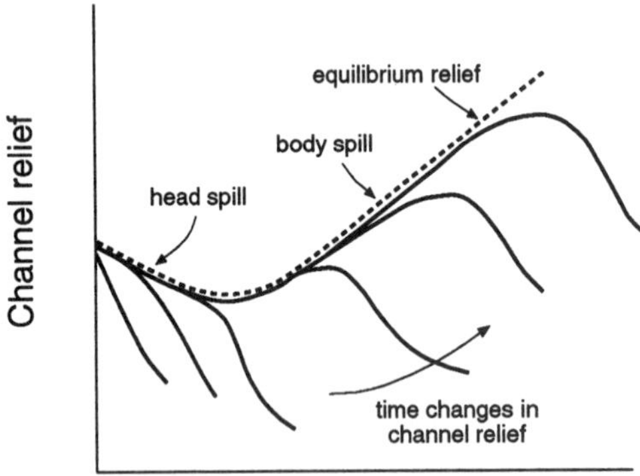

Figure 2.2 (b). Graph to show the evolution of channel relief in time, towards the expected equilibrium relief. Redrawn after Komar (1969).

over levees 900 m above the valley floor, and had velocities <1 m s^{-1} (Stow & Bowen 1980).

Menard (1955) introduced the concept of bankfull flow in leveed submarine channels, and invoked the effect of the Coriolis force in creating asymmetric levees. The amount of cross-flow gradient produced by the Coriolis force also depends on the type of turbidity current which exerts a greatest deflecting force on slow muddy flows, and results in the development of asymmetric levees and also controls the location of spillover points (Normark & Piper 1991).

Some modern submarine channels on mature passive-margin fans show highly sinuous fan channels, with evidence of breached levees on sharp bends (e.g. the Indus, Mississippi and Amazon Fan channels). Spillover points develop where flows breach levees on sharp bends, and where channel gradients decrease to cause an increase in flow thickness. If the spillover flow is sandy (i.e. commonly a large breach of the levee), then the sheet spillover flow may accelerate and erode a new channel (e.g. the Var Fan, Normark & Piper 1991). If the flow is muddy, however, then the spillover flow will decelerate, resulting in more sheet-like deposition of mud (e.g. turbidity current II on the Navy Fan, Piper & Normark 1983).

Flow-stripping processes can occur on sharp channel bends, where momentum changes cause the upper part of the flow to breach the levee. This leads to a rapid reduction in flow velocity, resulting in deposition of at least part of the coarser fraction from the lower part of a flow (Normark & Piper 1991). Indeed, flow stripping *sensu lato* may occur wherever deposition occurs along a turbidity current pathway such that the finer-grained suspension continues after the sandy grain-size population is dumped, e.g. over intra-channel bars and ledges, over large-scale erosional scours along a channel reach (mega-flutes, fault scarps, etc.), over levee collapse slides and within intraslope basins along channel paths.

The amount of flow stripping depends on the competence and capacity, magnitude of velocity change, grain-size distribution throughout a flow, and the momentum of turbidity currents with respect to channel depths. Large flows that are an order of magnitude thicker than the height of the levees, are much less affected by the constraints of a channel, and therefore flow stripping has less effect. Flow stripping is also insignificant for flows that are considerably smaller than the channel confines. Intermediate-sized flows, however, (e.g. flows twice as thick as the channel depth) may be strongly affected by flow-stripping processes. The effect of different flow characteristics on flow processes and deposit morphology is illustrated in Figure 2.3.

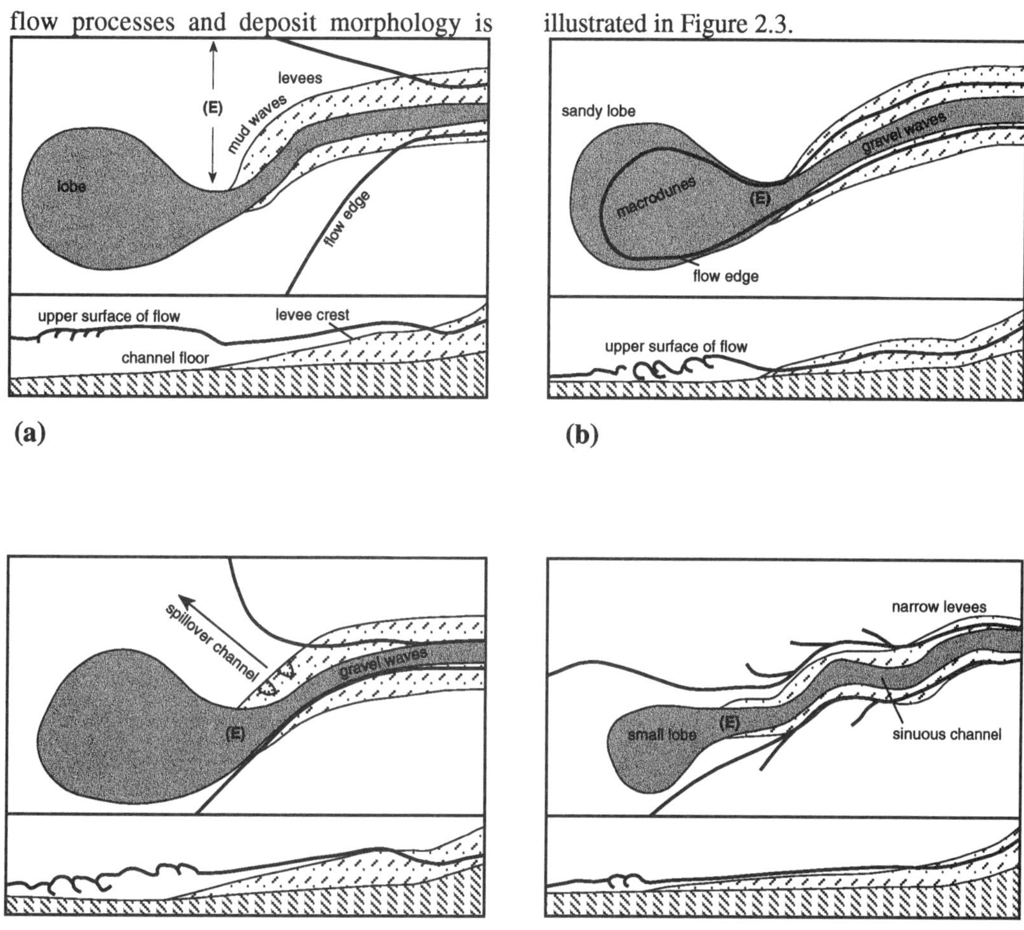

Figure 2.3. Flow paths and flow profiles for different types of turbidity currents through a simple canyon/leveed channel/lobe fan system. Spillover points and zones of water entrainment (E) are marked. (a) Large muddy flow. (b) Large sand-rich flow. (c) Large mixed sediment flow. (d) Series of repeated small muddy flows. Redrawn after Normark & Piper (1991).

Chapter 3

Architectural elements

Introduction

Deep-water deposits may be defined in terms of "architectural elements" (cf. fluvial models of Miall 1985). In fluvial sedimentology, Miall's (1985) "architectural elements" are characterised by facies associations and three-dimensional geometry (including orientation). The identification of depositional geometry requires the classification of a hierarchy of bounding surfaces, an approach now widely accepted for fluvial and eolian deposits. Using an hierarchy of bounding surfaces, together with facies analysis and geometry, we show how "architectural element analysis" may be used as an aid in the interpretation of deep-water systems, and their depositional history.

For fluvial sediments, architectural element analysis, both *sensu lato* and *sensu stricto*, has proved a useful tool for describing depositional units and modelling ancient channel deposits (e.g. Friend 1983, Allen 1983, Miall 1985, Bristow & Best 1993). Miall (1989) suggested the use of this technique for the interpretation of turbidite systems, and demonstrated its efficacy over more traditional vertical sequence analysis. A development of these techniques, in particular its application to smaller-scale features of turbidite systems, is described in Pickering *et al.* (1995b). Miall (1985) defined an architectural element as a *"lithosome characterised by its geometry, facies composition, and scale, and represents a*

particular process or suite of processes occurring within a depositional system" (Figure 3.1). The recognition of architectural geometries inherently involves the identification of bounding surfaces, and surface hierarchy. Hierarchical schemes have been implemented by many authors working on eolian and fluvial deposits (e.g. McKee & Weir 1953, Brookfield 1977, Allen 1983, Miall 1985). Allen's (1983) hierarchy of bounding surfaces in fluvial sediments, developed from his study of the Devonian Brownstones of the Welsh Borders, is commonly used in the description of fluvial deposits.

The concept of elements in turbidite systems was introduced by Mutti and Normark (1987) and is equally viable for modern and ancient settings: the elements recognised include large-scale features only, such as channels, overbank deposits, and lobes, together with scours which range across several orders of magnitude in scale. These are identified by facies associations and a classification of the hierarchy of events. Depositional and erosional features may also be identified at smaller scales within turbidite systems, including, for example, similar elements to those defined by Miall (1985) in fluvial sediments.

The characterisation of outcrops of ancient deposits into their component elements, on a variety of scales, will help unravel complex

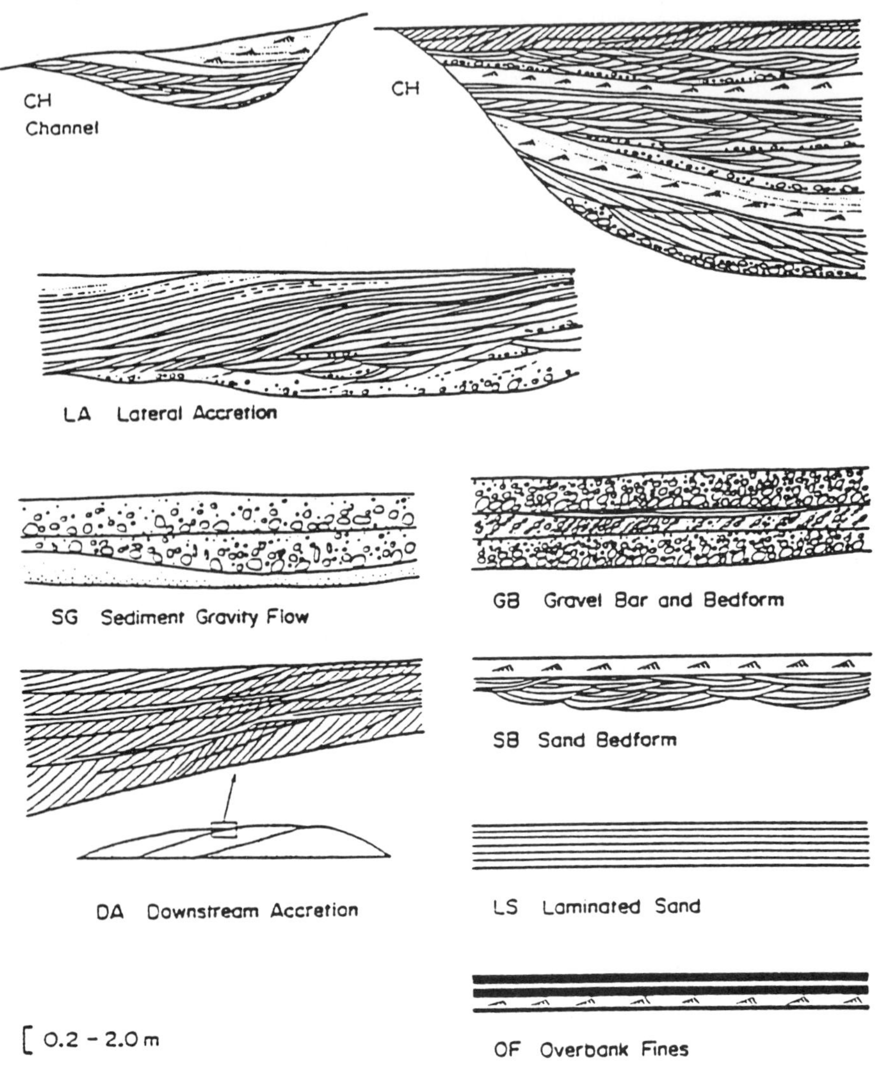

Figure 3.1. Architectural elements in fluvial deposits after Miall (1985).

depositional histories, and facilitate an understanding of the development of growth stages within turbidite systems. Not all sedimentary features of ancient deep-water deposits, however, can be expressed by Miall's (1985) architectural element scheme, for example sediment slides, mud mounds, and scour-and-fill features (on a variety of scales), and unlike Mutti and Normark's (1987) turbidite element scheme, the fluvial scheme is restricted to the description of ancient deposits. A more methodical approach to this type of analysis is required, and therefore it is proposed here to describe architectural elements, characterised by various facies and scales, from both modern and ancient environments.

Element analysis scheme

Depositional bodies comprise architectural elements, defined between bounding surfaces which may be erosional or conformable. Pickering *et al.* (1995b) reserve the term "architectural element" for the interpretative characterisation of a sedimentary feature defined by geometry (including orientation), scale, and facies. An appropriate measure of scale may be attained from identifying the hierarchy of the bounding surfaces defining the architectural element. The element analysis scheme may be used to describe depositional and erosional features of both modern and ancient systems (e.g. a gravel bar element may be a characteristic element in either a modern or ancient submarine channel deposit). Figure 3.2 shows the philosophical framework used to classify and interpret modern and ancient deep-marine environments.

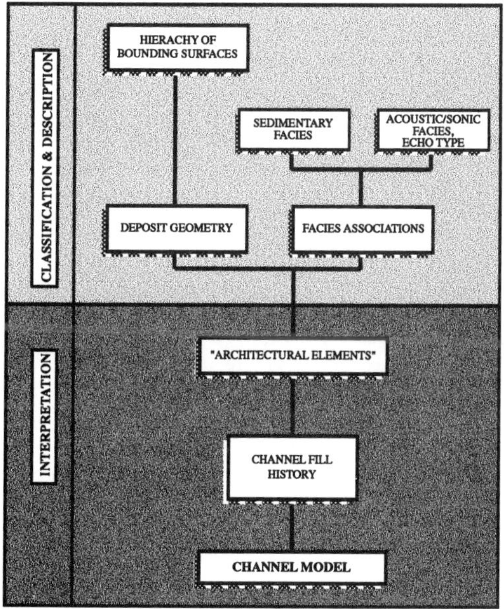

Figure 3.2. Architectural characterisation of modern and ancient depositional bodies in deep-water environments. After Pickering *et al.* (1995b).

The approach to depositional characterisation of deep-water deposits is as follows:- a bounding surface hierarchy scheme as applied in fluvial sedimentology allows deposits to be delineated at various scales into their architectural elements. The architectural elements are then classified using facies and "architectural geometry" classification schemes. Facies classifications for deep-water deposits exist in the literature (e.g. Mutti and Ricci Lucchi 1975, Pickering *et al.* 1986a, 1989, Ghibaudo 1992).

Architectural geometry

Architectural geometry is defined so as to be applicable over a wide range of scales (at least on scales greater than metres), but is not scale independent, and includes most facies. A classification of two-dimensional architectural geometries of what are in fact three-dimensional sedimentary features is preferred since the majority of geological observations in both outcrop and seafloor observations, are recorded in this manner. A three-dimensional geometrical classification is possible but its applicability would be limited, since very few case studies include such data quality. Without the third dimension, however, many features may appear similar, e.g. a large scour may have aspect ratios and a sedimentary fill identical to that of relatively small channels: their resolution and differentiation requires three dimensions. As with the application of any descriptive classification scheme, there are inherent ambiguities which should be appreciated.

Classifications of erosional-depositional two-dimensional architectural geometry both in section and in plan, are shown in Figures 3.3 and 3.4, respectively. These are, in section: channel, sheet, lens, sigmoid/inclined, irregular, waveform, and scour/scour-and-fill. In plan, the following elements are recognised: channel, nested channels, mound/lobes, nested mounds, irregular, bedform field, and scour. It should be noted that the "channel" is an architectural geometry within which all the other geometries may be found.

Although this classification of geometric elements is mainly applicable over a large range of scales (but not scale independent), the limitations of remote sensing of modern systems and subsurface ancient systems will introduce an unavoidable scale dependency into any geometrical classification (Normark *et al.* 1979).

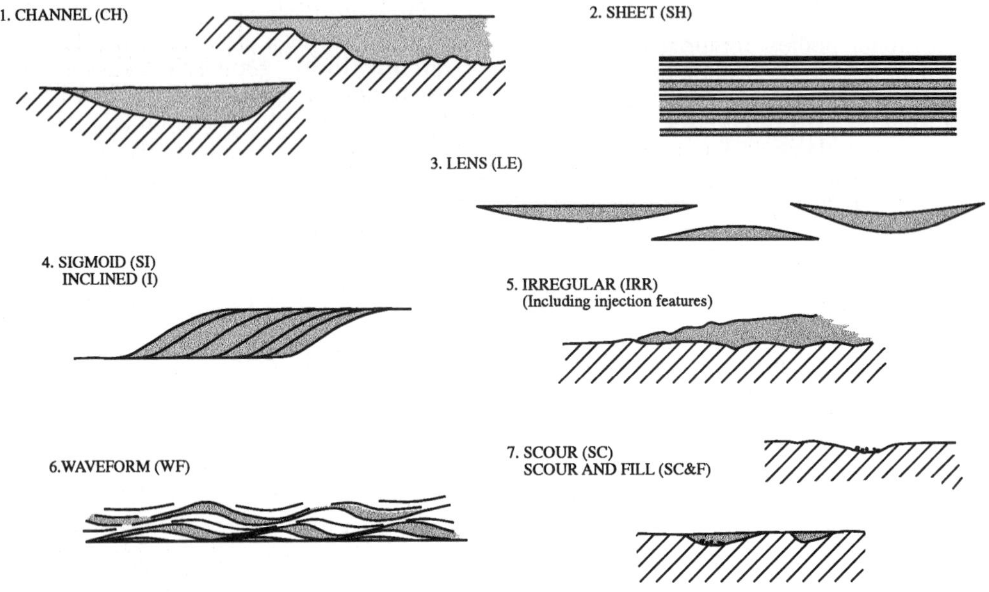

Figure 3.3. 2-Dimensional sectional classification of deep-water architectural geometry (scale independent). After Pickering *et al.* (1995b).

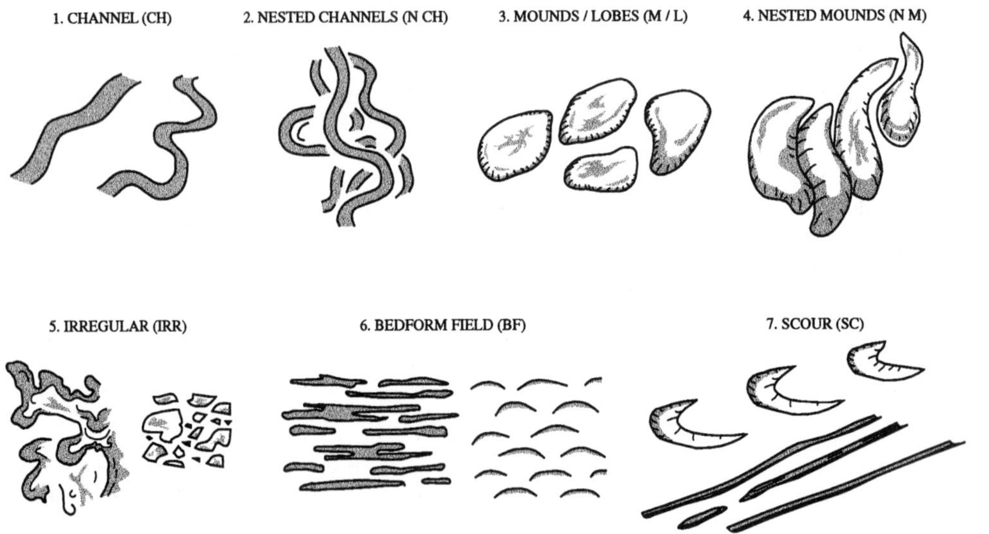

Figure 3.4. 2-Dimensional planform classification of deep-water architectural geometry (scale independent). After Pickering *et al.* (1995b).

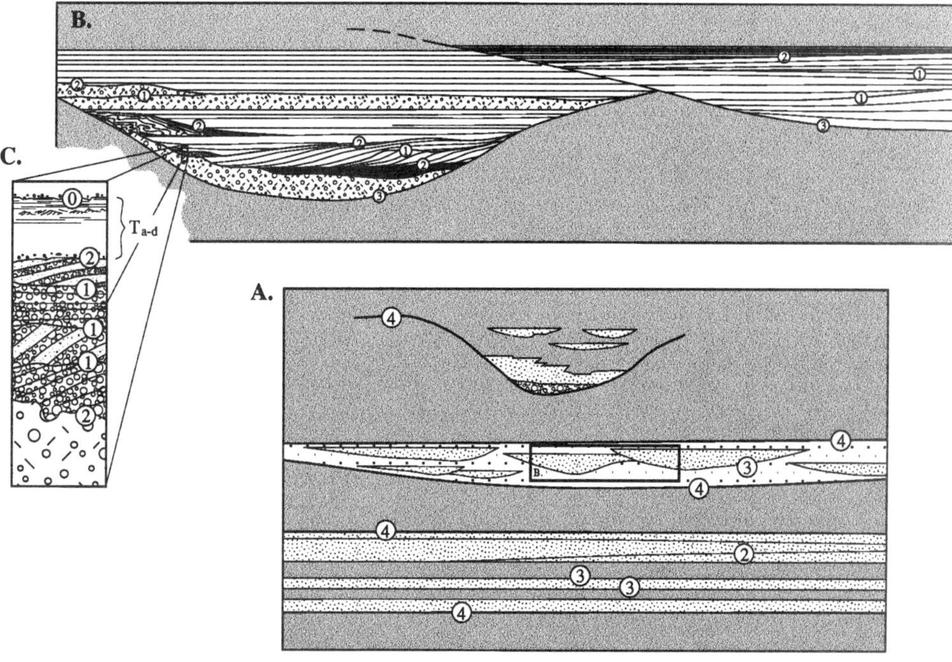

Figure 3.5. Schematic illustration of bounding surface hierarchy in outcrops of ancient deep-water deposits. After Pickering *et al.* (1995b).

Bounding surfaces

Bounding surfaces can be classified by type and order, both in cross-sections, and in plan. To date, geologists working in the ancient rock record have focused on the two-dimensional, cross-sectional, form of bounding surfaces. Allen (1983) recognised four types of bounding surface in the fluvial sediments of the Devonian Brownstones of the Welsh Borders: concordant non-erosional (normal bedding), discordant non-erosional, concordant erosional, and discordant erosional contacts. All such contacts can be found in turbidite systems.

Classifications of bounding surface hierarchy can also be readily applied to the submarine environment. Allen's (1983) classification is chosen here because of its simplicity, although fourth, fifth and sixth orders have been added to include basin-scale heterogeneity (similar to the scheme of Miall 1985, 1989). Figure 3.5 shows schematic sections of deep-water deposits, including a submarine channel, showing the bounding surface hierarchies and how the classification of these surfaces can easily be applied to sectional geometries of deep-water deposits (Pickering *et al.* 1995). The normal concordant bedding contacts between strata and laminae form the zeroth-order of bounding surface. *First-order surfaces* bound, for example, packages of cross-bedding sets or concordant bedding. Typically, these surfaces are erosional (concordant or discordant). *Second-order surfaces* bound units delineated by first-order surfaces to form distinct sedimentary complexes of genetically related facies and/or palaeocurrent directions. These complexes are equivalent to the "storeys" of Friend *et al.* (1979). *Third-order surfaces* are major erosional features dividing groupings of complexes (as delineated by second-order bounding surfaces). These units commonly are informally referred to in the literature as depositional bodies. *Fourth-order surfaces* have been added to express erosional contacts which can range up to a basin-wide scale, and define, for example, groups of channels and palaeovalleys and are equivalent to Miall's (1985) sixth-order surfaces, and separate Mutti and Normark's (1987) "stages of growth" within an individual deep-marine system (their third-order of physical scale). Mappable stratigraphic units, such as members or sub-members, are bound by these fourth-order surfaces (Miall 1989). *Fifth-order surfaces* define individual fan systems, and are equivalent to surfaces defining Mutti and Normark's (1987) second-order of physical scale. Finally sixth-order surfaces delineate basin-fill sequences and supergroups. These bodies are equivalent to Mutti and Normark's (1987) first-order of physical scale.

It should be noted that the appropriate hierarchy may differ between systems, simply because some turbidite systems are more punctuated into stages and sub-stages than others. In this respect, any hierarchy that is developed for one system should not be assumed a priori to be comparable to another.

Bounding surfaces seen in plan view are less easily defined by current classification schemes, but should nevertheless be used to help characterise architectural elements. Identification of such bounding surfaces, and bounding surface hierarchy, requires some interpretation of the element that is delineated by the bounding surface. For example, a major channel element seen

---

Class
Group
  Facies   See text for detailed descriptions

A Gravels, muddy gravels, gravelly muds, pebbly sands, ≥5% gravel
  A1 Disorganized gravels, muddy gravels, gravelly muds and pebbly sands
    A1.1 Disorganized gravel
    A1.2 Disorganized muddy gravel
    A1.3 Disorganized gravelly mud
    A1.4 Disorganized pebbly sand
  A2 Organized gravels and pebbly sands
    A2.1 Stratified gravel
    A2.2 Inversely graded gravel
    A2.3 Normally graded gravel
    A2.4 Graded-stratified gravel
    A2.5 Stratified pebbly sand
    A2.6 Inversely graded pebbly sand
    A2.7 Normally graded pebbly sand
    A2.8 Graded-stratified pebbly sand

B Sands, ≥80% sand grade, <5% pebble grade
  B1 Disorganized sands
    B1.1 Thick/medium-bedded, disorganized sands
    B1.2 Thin-bedded, coarse grained sands
  B2 Organized sands
    B2.1 Parallel-stratified sands
    B2.2 Cross-stratified sands

C Sand-mud couplets and muddy sands, 20–80% sand grade, <80% mud grade (mostly silt)
  C1 Disorganized muddy sands
    C1.1 Poorly sorted muddy sands
    C1.2 Mottled muddy sands
  C2 Organized sand-mud couplets
    C2.1 Very thick/thick-bedded sand-mud couplets
    C2.2 Medium bedded sand-mud couplets
    C2.3' Thin-bedded sand-mud couplets
    C2.4 Very thick/thick-bedded, mud-dominated, sand-mud couplets

D Silts, silty muds, and silt-mud couplets, >80% mud, ≥40% silt, 0–20% sand
  D1 Disorganized silts and silty muds
    D1.1 Structureless silts
    D1.2 Muddy silts
    D1.3 Mottled silt and mud
  D2 Organized silts and muddy silts
    D2.1 Graded-stratified silt
    D2.2 Thick irregular silt and mud laminae
    D2.3 Thin regular silt and mud laminae

E ≥95% Mud grade, <40% silt grade, <5% sand and coarser, ≤25% biogenics
  E1 Disorganized muds and clays
    E1.1 Structureless muds
    E1.2 Varicoloured muds
    E1.3 Mottled muds
  E2 Organized muds
    E2.1 Graded muds
    E2.2 Laminated muds and clays

F Chaotic deposits
  F1 Exotic clasts
    F1.1 Rubble
    F1.2 Dropstones and isolated ejecta
  F2 Contorted/disturbed strata
    F2.1 Coherent folded and contorted strata
    F2.2 Brecciated and balled strata

G Biogenic oozes (>75% biogenics), muddy oozes (50–75% biogenics), biogenic mud (25–50% biogenics) and chemogenic sediments, <5% terrigenous sand and gravel
  G1 Biogenic oozes and muddy oozes
    G1.1 Biogenic ooze
    G1.2 Muddy ooze
  G2 Biogenic muds
    G2.1 Biogenic mud
  G3 Chemogenic sediments

Table 3.1. Facies classes of Pickering *et al.* (1986a, 1989)

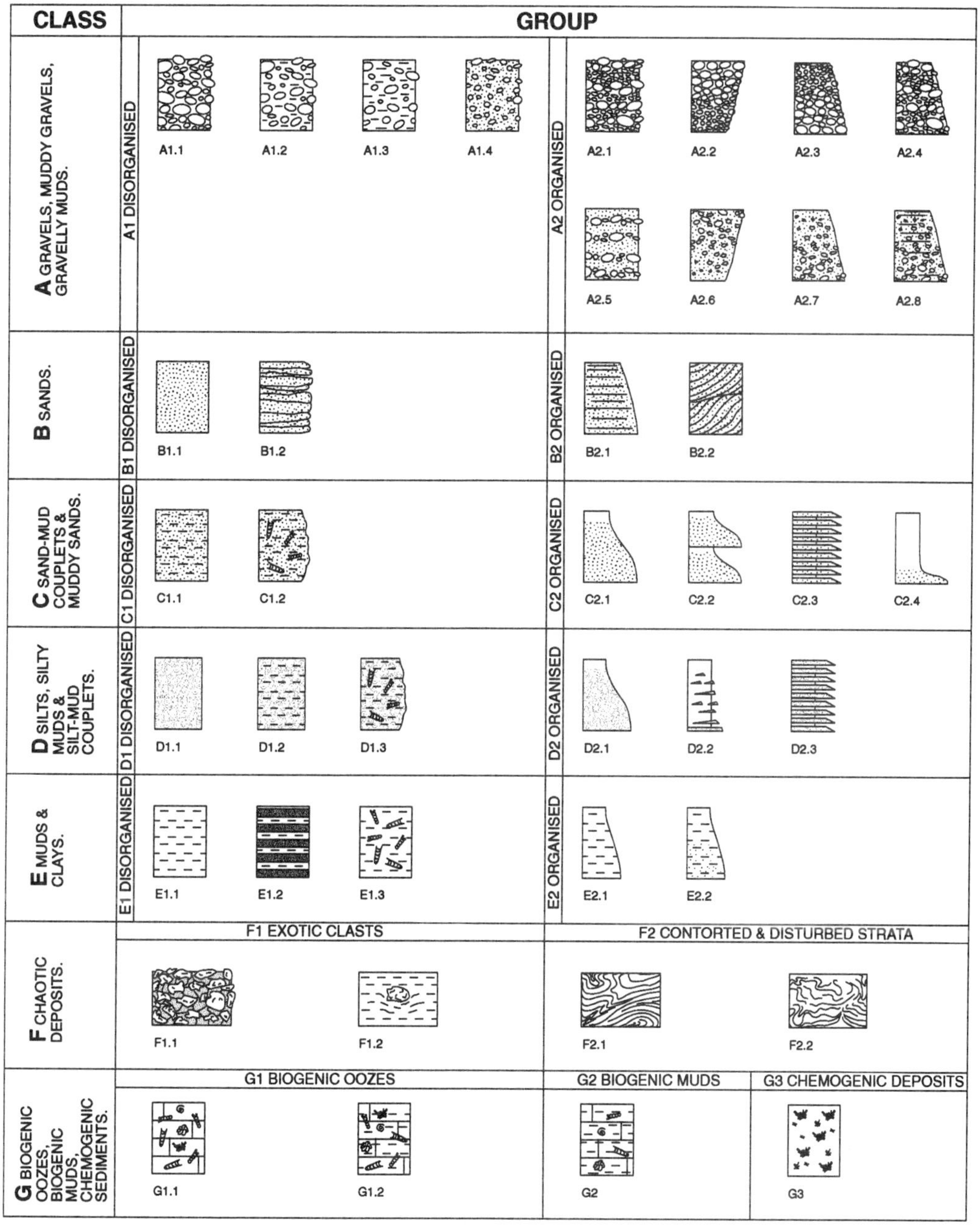

Figure 3.6. Facies classification scheme for deep-water sediments (after Pickering et al. 1986a, 1989, 1995b).

on sidescan sonar images will be defined by third-order bounding surfaces, using the similar diagnostic criteria as adopted for defining sectional bounding surfaces (see above).

Facies

Various facies schemes for deep-water sediments exist in the literature (e.g. Mutti & Ricci Lucchi 1972, 1975, Mutti 1977, 1992, Pickering *et al.* 1986a, 1989, Ghibaudo 1992). Here, we adopt the scheme of Pickering *et al.* (1986a, 1989) (Figure 3.6, Table 3.1). In systems where only core samples are available, the mapping of sedimentary features on the seafloor or subsurface is not possible, but where such information exists the sedimentary facies can be described as mentioned above. In the absence of such information, as is the case in many modern turbidite systems, sonar facies can be characterised but with equivocal interpretations as to their true sedimentological character. For planform analysis, acoustic facies from sidescan sonar images may simply be classified as high, moderate or low backscatter facies, with classifications of seismic facies in seismic profiles for sectional analysis already available (e.g. Damuth 1980, Droz & Bellaiche 1985, Pratson & Laine 1989).

Application of scheme

Once the hierarchy of bounding surfaces for modern and ancient systems is established, the geometry can be classified using the scheme proposed here. Architectural geometry bound by at least third-order surfaces is defined as third-order elements; second-order bounding surfaces define second order elements, etcetera. Sedimentary or acoustic facies can be assigned to the various orders of architectural geometries. These steps lead to a descriptive classification of the sedimentary feature. Using this description facilitates the interpretative characterisation of the feature by assigning to it an "architectural element". For example, a pebbly sandstone and clast-supported conglomerate second-order lens element within a submarine canyon, may be interpreted as a second-order gravel bar architectural element. Architectural elements are defined therefore by geometry (including orientation), order of bounding surface, and facies.

The purpose of a scheme such as the one adopted here is to permit the comparison of features between different deep-water systems. Furthermore, any formal classification scheme, with its attendant application, encourages a more accurate, less ambiguous, use of terminology at comparable scales.

Architectural element analysis as a tool for comparing modern and ancient systems

The description of deep-water sedimentary environments inevitably leads to comparisons being made with other ostensibly similar systems. Careful consideration should be given to the appropriateness of selected comparisons between modern and ancient deep-water systems. In the past, classifications of deep-water systems, and their tectono-sedimentary settings, have been at least implicitly assumed to compare similar features from modern and ancient systems (e.g. Mutti & Normark 1987, 1991, Stow 1986, Shanmugam & Moiola 1988). From the following discussion, we are sceptical that many of these modern and ancient comparisons are valid. The philosophy adopted for the schemes presented here is not concerned with comparing fan systems, type of basin, etcetera, but rather concentrates on describing the architectural elements within deep-water systems. Comparisons can then be made between similar architectural elements, or suites of elements.

Comparisons between geological features in deep-water systems must be made at similar spatial and temporal scales (Mutti & Normark 1987). In the scheme outlined here, the importance of an enveloping bounding surface, or its hierarchy with respect to other surfaces, is used as an appropriate measure of scale. Care must still be exercised, however, when comparing architectural elements of the same order, e.g. third-order channel elements may range considerably in size, but the importance of the bounding surface is only that it represents changes in depositional pattern, the interpretation that similar processes caused these changes is not justified without additional evidence.

The way in which geological and oceanographic observations are made for ancient and modern deep-water systems commonly results in a lack of common datasets. Field observations of ancient rocks provide detailed description of facies, facies-associations, vertical sequences, and identification of meso- and micro-scale sedimentary structures. These observations are not commonly available in any synthesised form to workers on modern deep-water systems. The mapping of modern deep-water environments is commonly restricted to the identification of larger-scale features. Long range side-scan sonar instruments (e.g. GLORIA) enable oceanographers and sedimentologists to map out in plan extensive areas of seafloor sedimentation. The best high-resolution of GLORIA, however, is >100 m: given that a typical field outcrop is commonly of this order of scale or smaller, the problems involved in identifying similar sized modern and ancient features become apparent. Detailed photogeological mapping of ancient systems may go some way to bridging this gap in observations (e.g. Sgavetti 1991).

Deep-towed side-scan instruments (e.g. TOBI, SEAMARC) give a more detailed view of the seafloor (resolution <20 m) and, although these instruments are not generally used for the mapping of extensive areas of deep-water systems, they can show architectural elements of the order of scale that could be identified from outcrop studies. Side-scan sonographs, most useful in modern marine studies, show plan-view sections of the sedimentation surface rarely obtainable in ancient outcrops at such scales. Good sectional data, however, can be obtained from outcrop studies, and the standard of sectional correlations between outcrops can not be matched using high-frequency seismic profilers, commonly used in the study of modern systems.

The importance of this scheme lies in the assumption that an architectural element defined from outcrop by facies, geometry and bounding surface hierarchy (which largely involves sectional analysis), can be compared with a similar architectural geometry defined by the same parameters in a modern deep-water system (from either plan or sectional analysis). For example, high-backscatter (on side-scan sonar), low-penetration (on high-frequency seismic profile), second-order bedform elements from a modern system, may be interpreted as relatively coarser grained second-order bedforms, and used for comparative studies and analogies with similar second-order ancient coarse-grained wave or bedform architectural geometry. Such a comparison does not necessarily lead to the interpretation that the two elements formed from similar processes.

The channel types of Normark (1970a), erosional, depositional and erosional-depositional or mixed, and their related fill deposits can be identified in both modern and ancient settings (see Figure 1.5). This submarine channel classification scheme is useful for modelling sediment supply, controlled by intra-basin, tectonic and eustatic sea-level controls. The Eocene Ainsa I and Ainsa II channel complexes in the southern-central Pyrenees, provide good examples of mixed erosional/depositional channels (Mutti & Normark 1991). In the northern North Sea Paleogene hydrocarbon province, good examples of slope and lower slope erosional-depositional channel systems are documented from the Paleocene Balder Formation in UK Quadrant 9 by Timbrell (1993), the Paleocene-Eocene Gryphon Field by Newman *et al.* (1993), and the Eocene Alba Field in UK Block 16/26 (Newton & Flanagan 1993).

29

Part II

Modern Systems

Chapter 4

Surveying and sampling seafloor sediments

Seismic profiles

Systematic mapping of the seafloor began in the late 1950's with the use of high-resolution seismic surveys (Heezen *et al.* 1959, Shepard *et al.* 1960, Menard 1964, van Andel & Shor 1964), which allowed the characterisation of large areas of sediment covered seafloor. Surveys such as these relied on the analysis of microphysiography and also reflectivity characteristics of the seafloor (collectively known as echo character). High-frequency seismic waves are required to obtain the high-resolution detail of the surface and near surface sediments. Commonly 3.5-kHz seismic records are used for this purpose which enable 1-2 m resolution at the sea-bed, and typically allow sufficient penetration into the seafloor sediments to reveal details of echo character. In addition 10-kHz, 12-kHz and 100-kHz seismic profiling surveys can be used for seafloor surveying, providing higher resolution of seafloor topography, with less acoustic penetration into the buried sediments. Acoustic facies mapping with 3.5-kHz seismic records was developed in the 1970's (see review by Damuth 1980).

The 3.5-kHz records are a useful tool for locating and identifying submarine channels, characterising sectional morphology, providing sufficient vertical resolution of the topographic features, and determining the echo character of the channel-fill, channel-margin and channel-levee sediments. Conventional, low frequency, multi-channel seismics have been routinely used to evaluate the vertical stacking of large submarine channel complexes on submarine fans (e.g. Weimer & Buffler 1988). This type of seismic survey allows greater penetration into fan sediments, but results in a loss of vertical resolution, something that is important for detailed channel characterisation.

Sidescan surveys

Ship-towed sidescan sonar instruments allow acoustic scanning of the seafloor. An array of transducers transmit a narrow fan-shaped beam of sonic pulses sideways from the sonar instrument and record the strength of the seafloor-reflected echoes that return. Most of the transmitted signal will be lost, but some "backscatter" will return to the sidescan instrument depending on the slope or the surface roughness of the seafloor. The backscatter intensities from the sea bed are recorded against time to produce sidescan sonographs. These techniques were first deployed in the 1960's for use in shallow water (see Belderson *et al.* 1972).

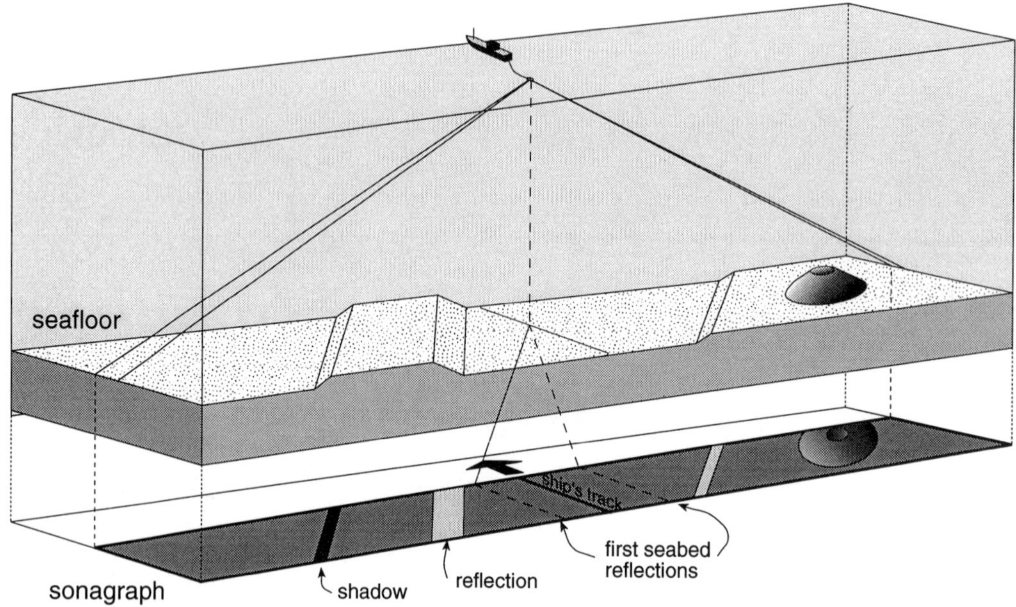

Figure 4.1. Schematic sketch to show the principle of sidescan sonar imaging of the seafloor (redrawn from Somers & Searle 1984)

The GLORIA (Geological Long Range Inclined Asdic) sidescan sonar devise, was developed by the Institute of Oceanographic Sciences, UK, to produce acoustic sonographs of the deep ocean seafloor. GLORIA operates at a frequency of 6.5-kHz, possesses a swath width up to 45 km, and can be towed at 8-10 knots, 50 m below sea-level (Sommers *et al.* 1978, Sommers & Searle 1984) (Figure 4.1). The sidescan resolution is 125 m along track and 50 m across track, roughly the size of a football pitch.

Although the GLORIA sidescan instrument has proved very useful in mapping large areas of the seafloor, its low resolution power constrains its effectiveness for more detailed examination of the seafloor. Deep-towed sidescan instruments enable higher frequency devices to be employed near the sea bed, and permit an increase in sidescan resolution power. TOBI (Towed Ocean Bottom Instrument) operates a 30-kHz sidescan sonar, and 7-kHz high resolution seismic profiler. It is typically "flown" about 400 m above the seafloor, and has a swath width of 6 km (3 km either side). The sidescan resolution ranges from 4 m along track and 7 m across track, to 42 m along track and 2 m across track in the

near and far ranges, respectively (Flewellen *et al.* 1993).

Other sidescan instruments have been used for deep-sea surveys. These include surface-towed systems such as the SeaMARC II (12-kHz, with a 10 km swath width), the Russian OKEAN long-range sidescan sonar (30 kHz), the Japanese IZANAGI, and deep-towed high resolution sidescan instruments such as SeaMARC I, the French SAR system (180 kHz) and Russian MAK system (100 kHz).

The geological causes of sonar backscatter are poorly known. Some of the factors which may be important in governing the strength of the backscattered signal include angle of incidence, seafloor topography, sediment lithology, surface roughness, surface hardness (i.e. the presence of "bedrock", semi-lithified or lithified hardgrounds and chemogenic iron- or magnesium rich layers at or near the surface) and the degree of inhomogeneity in the upper few metres of sediment. In one "ground truthing" study on the distal lobe area of the Monterey Fan, known as the "Fingers Area", sediment cores were taken over an area surveyed by GLORIA in an attempt to correlate sediment acoustic properties with backscatter strength (Gardner *et al.* 1991). The general results from this study show that high backscatter areas are associated with areas of slightly higher relief, and areas where less penetration is seen on 3.5 kHz profiles. It was qualitatively shown that the amount of inhomogeneities in the top few metres of sediment have a strong control on backscatter intensities (Gardner *et al.* 1991). Insonified sediment layers with different acoustic impedance properties (e.g. interbedded silt clay and thin sands), produce complex constructive and destructive interference patterns, which may lead to a greater amount of backscatter intensity than a thick homogenous sediment layer. In the "Fingers Area" of the Monterey lobe, areas of thick-bedded sands generally correlated with low backscatter areas (Gardner *et al.* 1991). These results are contrary to the conventional interpretation of sidescan backscatter seen in other areas of the Monterey Fan (Gardner *et al.* 1991), where high backscatter correlates with generally more sandy areas, and low backscatter with muddy areas. The study also concluded that surface roughness (in the form of small-scale bedforms and rock clasts seen from bottom photographs) is not a significant contribution to sonar backscatter.

The extent of penetration (of the sidescan signal into the sediment) which significantly contributes to the intensity of the backscattered signal is still unknown. For GLORIA sidescan at least the upper 3 m of the seabed are sampled (Gardner *et al.* 1991). Higher frequency sidescan instruments will have less penetration. The significance of the depth of sediment sampled in producing the backscatter intensity becomes apparent when attempting to correlate high and low resolution sidescan surveys from the same area. For example high-backscatter sediment wave fields observed on a GLORIA survey (Masson *et al.* 1991) were not observed as high backscatter area on a TOBI survey of the same site (unpublished data from Discovery cruise 205, West African Continental Rise, 1993). Without more "ground truthing" in sidescan survey areas, lithological interpretations of backscatter acoustic facies are very uncertain.

Swath bathymetry

Multibeam echo sounders are commonly deployed to obtain detailed bathymetry maps of the seafloor. Single beam echo sounders have since become obsolete for marine geology surveys of seafloor bathymetry. Seabeam was initially developed by the US Navy, but is now owned and greatly improved by the French. Seabeam, mounted in the ship's hull, has a swath width 3/4 that of the water depth (Renard & Allenou 1979). It has been successfully deployed on the Laurentian Fan (Hughes Clarke *et al.* 1990) and the Rhone Canyon and Upper Fan (Droz & Bellaiche 1985). Other swath bathymetry echo sounders include the German Hydrosweep and the Norwegian Simrad, which have been most commonly used in the last five years.

Sediment coring

The most commonly employed method of obtaining cores in deep-marine sedimentology is the use of piston corers and gravity corers. These methods however, are not successful in retrieving sandy or conglomeratic material, as they are unable to penetrate deep into an area of seafloor consisting of compacted coarse-grained material, and commonly any sand that is captured in the core is lost out of the bottom of the core during ascent to the surface.

Box corers (50 cm^2 area and maximum core length 80 cm), and the Kastonlot corer (20 cm^2 area and core length 4 m max.) have been used more successfully to core sandy seafloors. Legs 96 and 155 of the Deep Sea Drilling Program have successfully retrieved deep-drilled cores from the Mississippi (Figure 4.2) and Amazon Fan channels, respectively (see Chapter 5). As well as providing sediment to be visually analysed, sediment cores can also be analysed using x-ray photographs and gamma-ray logging techniques.

Bottom photographs, videos and submersible dives

Marine geologists and oceanographers commonly use bottom photographs and videos to obtain visual images of the seafloor, for example video pictures of gravel waves on the Laurentian Fan valley floor (Hughes-Clarke *et al.* 1990), and photographs of deposits on the floor of the Var Canyon (Malinverno *et al.* 1988). Deep-water submersibles have also been used (e.g. the Pisces and Alvin submersibles).

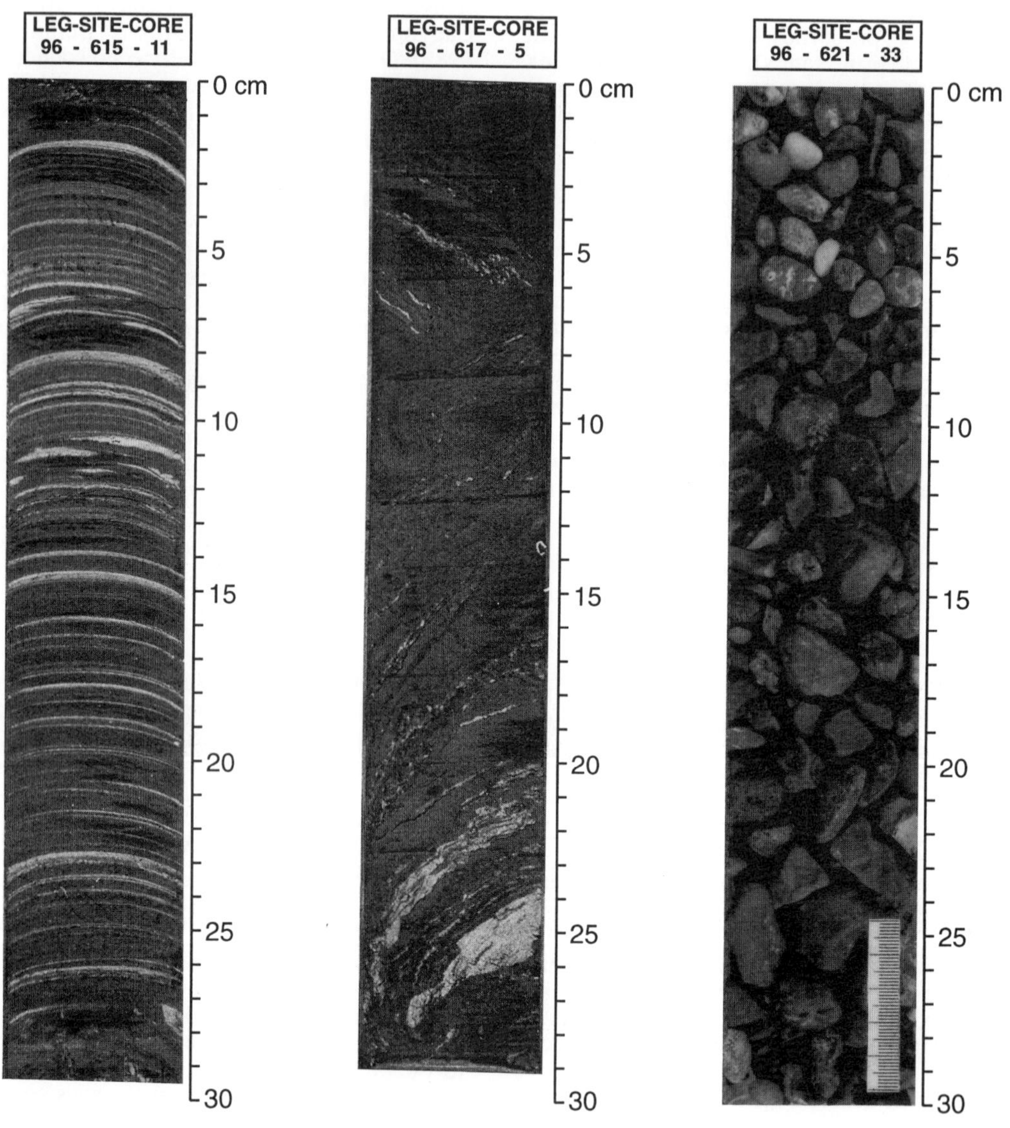

Figure 4.2. Core photographs from the Mississippi Fan channel, showing representative facies from the lower middle fan channel sites (DSDP Sites 615, 623, 624), middle fan channel (DSDP Sites 621, 622), and overbank-levee (DSDP Site 617). See Pickering *et al.* (1986b).

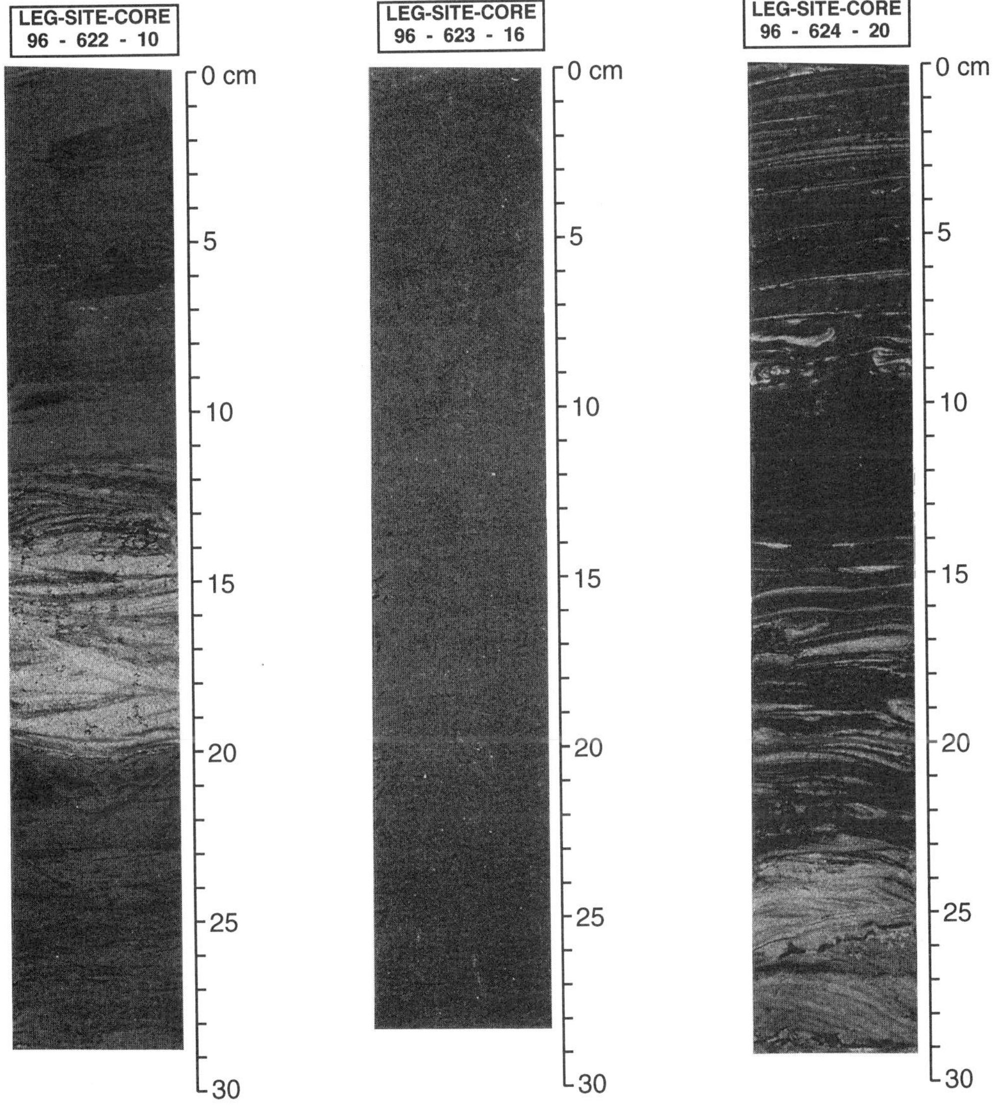

LEG-SITE-CORE
96 - 622 - 10
0 cm
5
10
15
20
25
30
LEG-SITE-CORE
96 - 623 - 16
0 cm
5
10
15
20
25
30
LEG-SITE-CORE
96 - 624 - 20
0 cm
5
10
15
20
25
30

Chapter 5

Modern channel systems

Introduction to dataset

The use of long-range sidescan sonar, swath bathymetry and high-frequency seismic profiles, allows the morphology of modern submarine channel-complexes to be mapped out in great detail. Flow processes may be inferred from morphological studies of submarine channels. Quantitative parameters can be measured directly from sidescan and seismic data, allowing quantitative analysis of channel complexes (see Chapter 6). In order to acquire such information, a large dataset of submarine channels has been studied, with an examination of the morphological and sedimentological characteristics of deep-marine channels from a variety of different basin settings. The channels used for quantitative analysis include those from active and passive margin settings, and are briefly described in the following case studies.

The GLORIA surveys from the EEZ-SCAN surveys (Exclusive Economic Zone), provide 200 nautical miles (370 km) of GLORIA data around the coastlines of the United States, published as a valuable data source in three Atlases (EEZ-Scan 84 Scientific Staff 1986, EEZ-Scan 85 Scientific Staff 1987, EEZ-Scan 87 Scientific Staff 1990). The West Coast of the USA contains deep-marine channels along an active margin, whereas the East Coast contains channels along a mature passive continental slope and rise. The EEZ-SCAN survey of the Gulf of Mexico provides detailed mapping of the Mississippi and DeSoto channels.

GLORIA surveys of the Aleutian Basin provide data of three major channel complexes sourced from volcanic islands along an active margin. These are termed the Umnak Channel, Bering Channel and the Pochnoi Channel (Kenyon & Millington 1995). GLORIA surveys have also allowed extensive mapping of the Indus and Amazon fan channels. The Rhône and Var channels in the western Mediterranean have been mapped from GLORIA data and detailed Seabeam surveys.

Monterey Channel, West Coast USA

Age:	Modern
Fan position:	Canyon to fan
Widths (km):	1.6, 2, 1.9, 2.8, 1.9, 1.875, 1.2, 0.9, 1.3, 0.4
Depths (m):	821, 884, 175, 188, 96, 47, 58, 60, 75, 32
Length:	300+ km
Aspect Ratios:	2, 2.3, 10.7, 15, 19.5, 40, 20.6, 15.6, 17.5, 11.7
Max. sinuosity:	1.29

The Monterey Channel is a moderately sinuous channel which extends 300 km across the upper and middle fan regions of the Monterey Fan (Normark *et al.* 1984, EEZ-SCAN 84 Scientific Staff 1986, 1988). The Monterey Fan is located off the west coast of Central California, and fed by two major canyon systems, the Monterey and the Ascension canyons located in the Monterey Bay area. The Ascension Canyon only extends back to the edge of the continental shelf, while the Monterey Canyon extends across the entire shelf, close to the shoreline and, therefore, is more active as a sediment conduit. The Monterey Canyon has remained active during the present marine high-stand, contrary to the relatively abandoned Ascension Canyon system (Normark 1985). The Monterey Fan probably developed during the late Oligocene or early Miocene (Normark *et al.* 1984).

GLORIA surveys of the fan surface have permitted the mapping of the slightly sinuous channel course, and show the development of braided distributary network on the active lobe areas (EEZ-SCAN 84 Scientific staff 1988, Gardner *et al.* 1991) (Figure 5.1). A high resolution side-scan sonar survey of the Monterey Channel, undertaken in 1990 using the TOBI deep-tow system, has delineated the sedimentary features of the present day channel-fill (Masson *et al.* 1995) (Figure 5.2). Figure 5.3 shows an image interpretation of a portion of the TOBI survey, and the main architectural features of the channel. The results of the survey show that this part of the channel is close to or in equilibrium between erosive and depositional processes, shown by the identification of both erosional and depositional architectural elements (Masson *et al.* 1995). The range of channel elements include asymmetric terraced walls, transverse step-downs in the channel floor which have the appearance of waterfalls, transverse and longitudinal bedforms. The bedforms are identified from bands of contrasting backscatter, and the longitudinal bedforms have a relief of 1-2 m seen on reflection profiles. The longitudinal bedforms preferentially develop in straight sections of the channel, and the transverse bedforms are common down stream from the "waterfalls" (Masson *et al.* 1995) (see Figure 5.3). The overbank features include sediment waves perpendicular or strongly oblique to the main channel, subsidiary channels and gullies formed from headwall erosion, scours, arcuate scarps and scattered blocks. The scours are 10 m to 1 km in diameter, can exceed depths of 10 m, and are comparable in size and appearance to those described by Shor *et al.* (1990) from the Grand Banks turbidity current (Masson *et al.* 1995). Some of the scours occur in groups immediately below channel bends (see Figure 5.3), suggesting that they possibly form from overbank process, similar to the flow stripping processes described by Piper & Normark (1983).

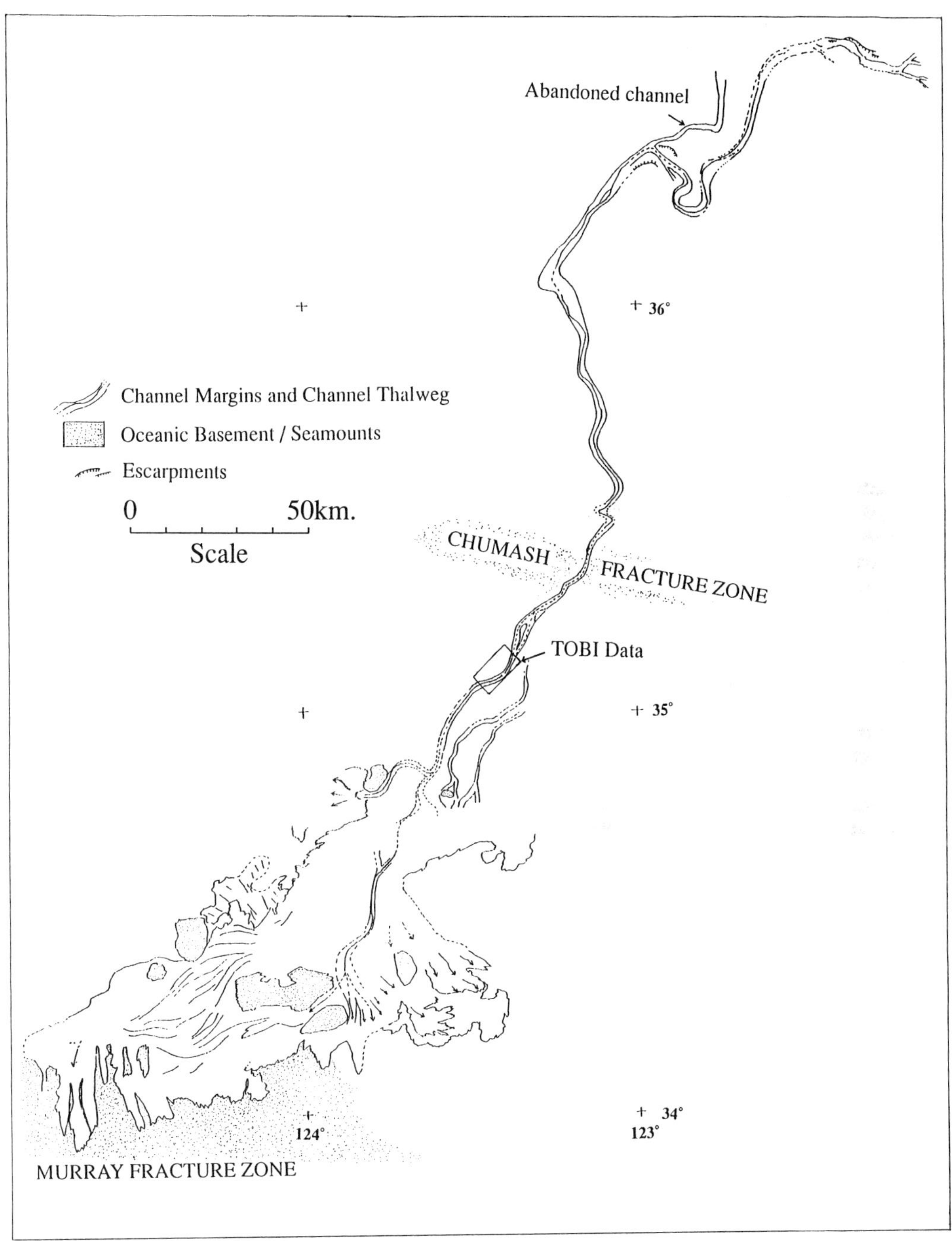

Figure 5.1. GLORIA interpretation of the Monterey Channel, West Coast USA.

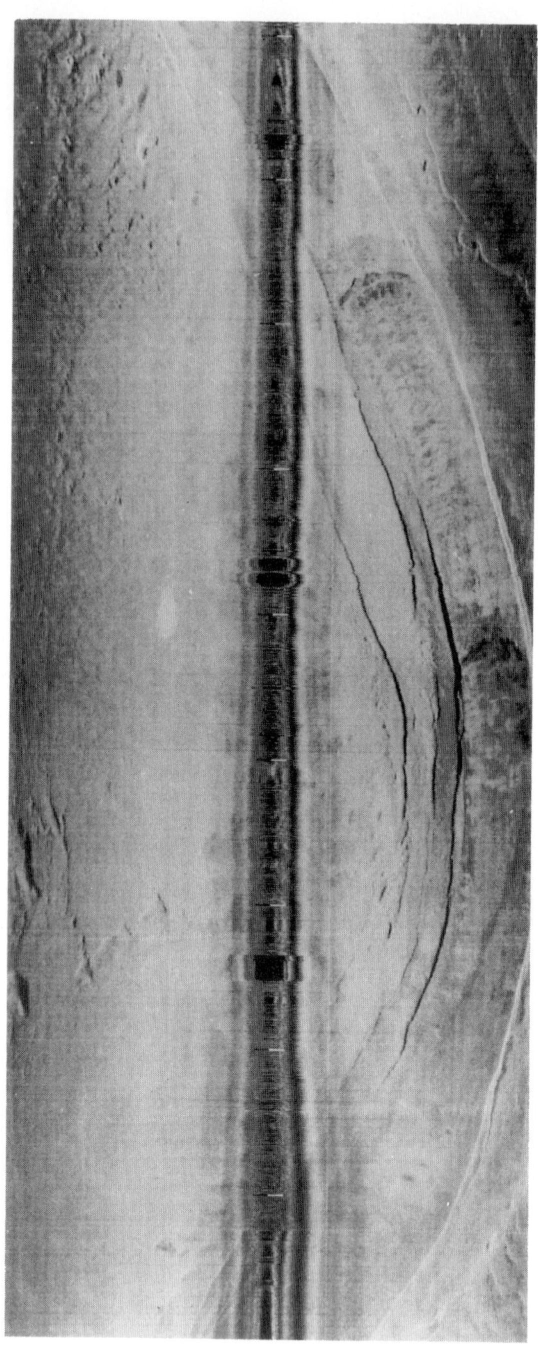

Figure 5.2. TOBI deep-tow sidescan sonar image of a low sinuosity fan channel on Monterey Fan, west coast USA (Masson *et al.* 1995).

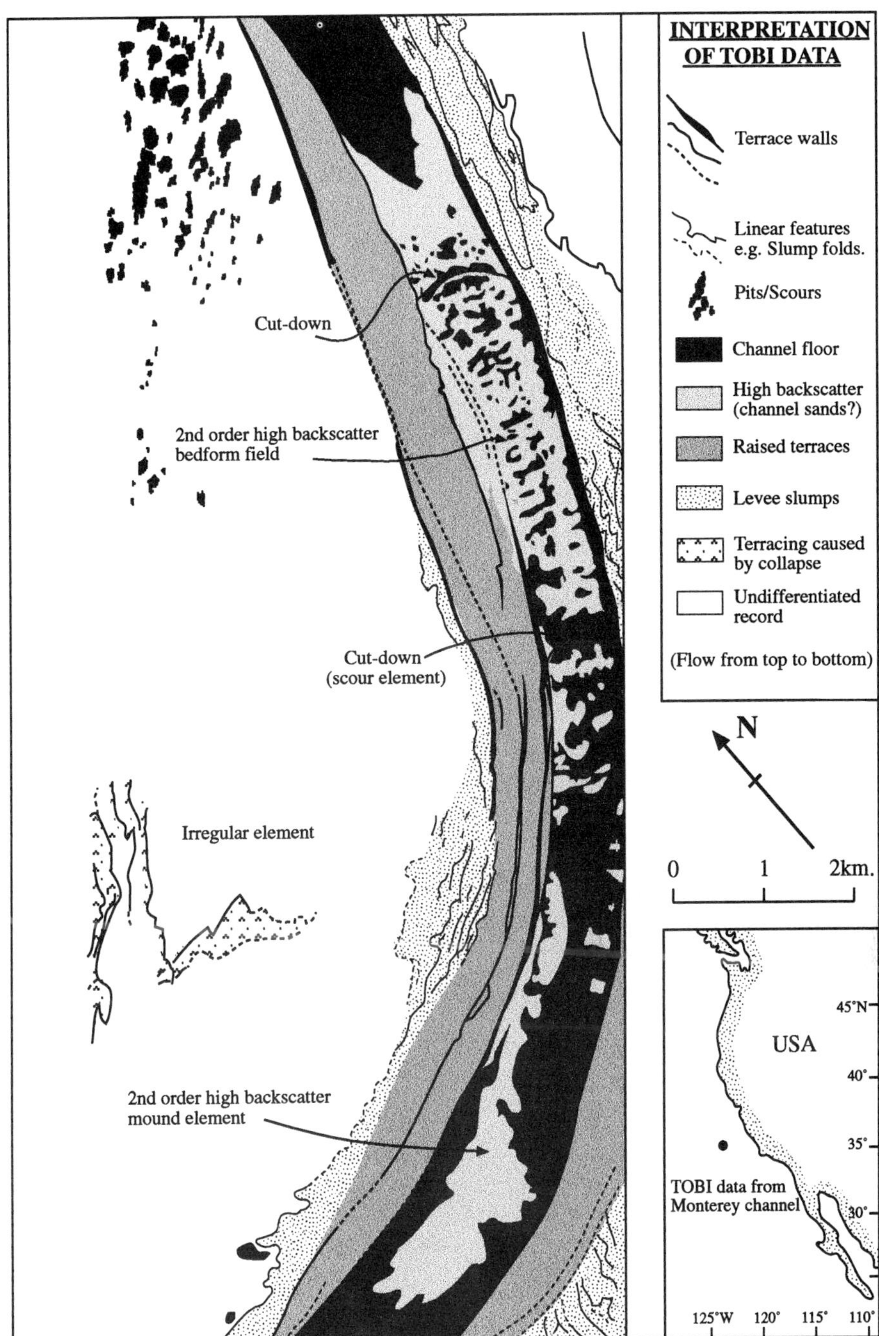

Figure 5.3. Interpretation of TOBI deep-tow sidescan sonar image of the Monterey fan-channel. Note the erosional steps and terraces, high-backscatter bedform field, and sediment slumps on the levees. Preferential sites of possible sand accumulation appear to include the high-backscatter bedforms and the intrachannel high-backscatter mound element immediately down-reach from the channel bend. Interpretation after Clark (1995).

Cascadia Channel, West Coast USA

Age:	Modern
Fan position:	Mid-ocean channel
Widths (km):	2.6, 1.5, 2.1, 1.5, 4.9, 5.6
Depths (m):	83, 107, 234, 285, 170, 165
Length:	2000+ km
Aspect Ratios:	32, 14, 9.1, 5.1, 28, 34
Max. sinuosity:	1.41

The Cascadia Channel, located off the Oregon-Washington continental slope, extends 2000 km across the Cascadia Basin (Figure 5.4). The basin is bound to the west by the Juan de Fuca Ridge, and to the south by the Blanco Fracture Zone, and has a southward dipping basin floor. The Cascadia Basin contains the Nitinat and Astoria fans, and major channel systems, the Cascadia and Vancouver channels. Numerous steep-sided incised canyons on the Washington continental margin, some extending up beyond the shelf break, supply sediment to the basin from along-shelf drift or direct fluvial input across the shelf during low sea-level periods (Hampton *et al.* 1989). The wedge of accreted sediment has formed "borderland" slope basins and ridges (Barnard 1978). Hampton *et al.* (1989) showed that accretionary tectonics have strongly controlled drainage patterns in the basin, commonly redirecting canyon courses or severing links with channels within the basin.

The Cascadia Channel is the main distributary arm of the Willapa-Cascadia-Vancouver-Juan de Fuca valley system, a tributary network of submarine channels. The main feeder channels are the "Cascadia Tributary" (Hampton *et al.* 1989), Willapa Channel, extending from the Willapa Canyon to the junction with the Cascadia Tributary, the Vancouver channel, and the Juan de Fuca Channel. The Juan de Fuca Channel is presently inactive, abandoned from changing drainage patterns resulting from the development of the accretionary prism (Hampton *et al.* 1989).

The Willapa Channel probably is the most important feeder channel, showing evidence on seismic sections to suggest that it is a relatively long lived feature. In contrast, the Cascadia Tributary may be a relatively young channel system (Griggs & Kulm 1973, Hampton *et al.* 1989). The correlation of the Mazama ash horizons from cores in the Cascadia Channel show evidence that turbidity currents are earthquake initiated (Adams 1990). The cores show that thirteen regularly spaced turbidity currents have occurred since the Mazama eruption (6845 ± 50 radiocarbon yr. B.P.), and it appears that turbidity currents occurred synchronously in the different tributary canyons, and coalesced to form single deposits within the main Cascadia Channel (Adams 1990).

The Cascadia Channel was formed in the Pliocene, at which time it probably followed a course located in the present-day Vancouver Seavalley (Griggs & Kulm 1973). The Vancouver Channel joins the Cascadia Channel as a "hanging valley", 60 m above the floor of the Cascadia Channel (Griggs & Kulm 1970). This is due to high-rates of downcutting within the Cascadia Channel, from recent (<6600 years) subsidence of the basin floor in the Blanco Fracture Zone to the south (Griggs & Kulm 1973, Embly 1985). The tectonic lowering of the channel has considerably affected the recent development and morphology of the Cascadia Channel. The upper reaches of the channel are aggradational with levees, whereas the lower reaches of the channel (south of 45.5°N) are in a state of erosional down-cutting and have no levees (Griggs & Kulm 1970, 1973). GLORIA sonographs show the development of linear terraces in this part of the Cascadia Channel, and the channel is widest (3.5 km) and deepest (300 m) at this point (Hampton *et al.* 1989). Hampton *et al.* (1989) suggested that terraces observed on sonographs from the Juan de Fuca and Vancouver channels may be accentuated by channel-floor sediment heterogeneity.

In summary, the course of the Cascadia Channel is strongly controlled by the tectonic configuration of the basin. Two major abrupt changes in channel direction take place due to the southward dipping basin floor, and where the channel encounters the Blanco Fracture Zone. The accreting continental slope also has played a role in the distribution of the drainage system, causing the abandonment and redirection of submarine channels throughout the evolution of the Cascadia Basin.

Figure 5.4. GLORIA interpretation of the Cascadia Channel and associated seafloor features.

Astoria Channel, West Coast USA

Age:	Modern
Fan position:	Canyon to fan
Widths (km):	2.9, 2.7
Depths (m):	165, 32
Length:	350+ km
Aspect Ratios:	17.5, 83
Max. sinuosity:	1.07

The Astoria Channel is part of the Astoria Fan, located in the Cascadia Basin. The Astoria Fan which developed during the Pleistocene, rests unconformably on Pliocene and Pleistocene aged abyssal plain turbidites (Kulm *et al.* 1973). The Astoria Fan is supplied by sediment from the Columbia River via the Astoria Canyon, and represents a good example of a point-sourced fan system (Nelson *et al.* 1970, Nelson 1985). The Astoria Fan is unconfined by basin topography, although the Cascadia Channel at its western margin, captures sediment from the Astoria Fan to limit its seaward growth (Nelson & Nilsen 1974). From the mouth of the Astoria Canyon the Astoria Channel can be traced across the entire length of the fan, until it intersects the "slope base fan valley" which runs along the base of the Oregon continental slope (Hampton *et al.* 1989).

The Astoria Channel is distinguished from the canyon by the presence of levees, and follows a moderately sinuous course across the fan (Hampton *et al.* 1989) (Figure 5.5). On the lower fan, the channel is difficult to observe from the GLORIA sonographs, and the coarse irregular backscatter pattern suggests the presence of a lower fan channel-distributary system (Hampton *et al.* 1989). Some of the distributary channels, seen as segments on the sonographs, may be isolated features, and could be remnant channels representing the course of previous positions of the main Astoria Channel (Hampton *et al.* 1989). The high silt and clay content of the Columbia River results in large turbidity currents being able to transport sediment to distal areas of the fan via the Astoria Channel (Nelson & Nilsen 1974, Nelson 1985). The outer fan, therefore, is relatively sand-rich, with the Deep Sea Drilling Project core (DSDP Site 174) showing an approximately 1:1 sand:mud ratio

in 284 m of outer fan sediments (Kulm *et al.* 1973). The Astoria Fan is efficient in sand transportation to its outer regions (Nelson & Nilsen 1974, Nelson 1985) and represents a high-efficiency system *sensu* Mutti (1979).

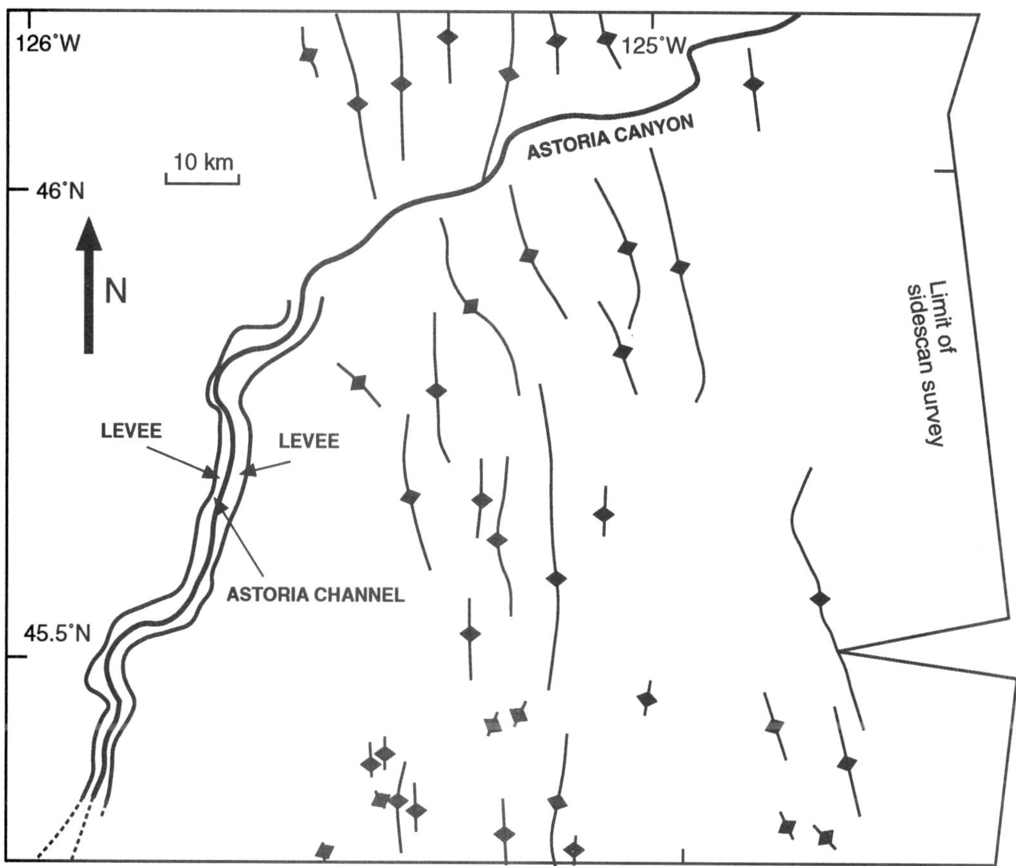

Figure 5.5. Gloria interpretation of the Astoria upper fan channel and accretionary prism of the west coast USA (Hampton *et al.* 1989).

Arguello Channel, West Coast USA

Age:	Modern
Fan position:	Canyon to fan
Widths (km):	1.5, 1.8, 1.1, 0.9, 0.6, 0.7
Depths (m):	302, 68, 166, 106, 69, 229
Length:	250 km
Aspect Ratios:	5, 26, 6.8, 8.8, 8.7, 2.9
Max. sinuosity:	1.54

The Arguello Channel lies south of the Monterey Channel, off Point Arguello, west coast of California. The channel is fed from a multiple canyon system (Figure 5.6) On the GLORIA sonographs the channel can be traced 250 km to a distal lobe infilling a localised depression in the oceanic crust. The channel mouth depositional lobe consists of an area of low backscatter surrounded by an area of high backscatter. The channel is difficult to trace in its middle reaches, and shows clear signs of multiple channel switching in the area in the recent past (see Figure 5.6). The channel then takes a 90° right-turn, due to the ocean floor topography. Beyond this point, the channel has a more sinuous form, of tight channel bends with small radii of curvature. The increase in sinuosity is attributed a thin covering of sediment over the oceanic crust allowing basement irregularities to control the channel course (EEZ-SCAN 84 Scientific Staff 1988).

Figure 5.6. GLORIA interpretation of the Arguello Channel, West Coast USA

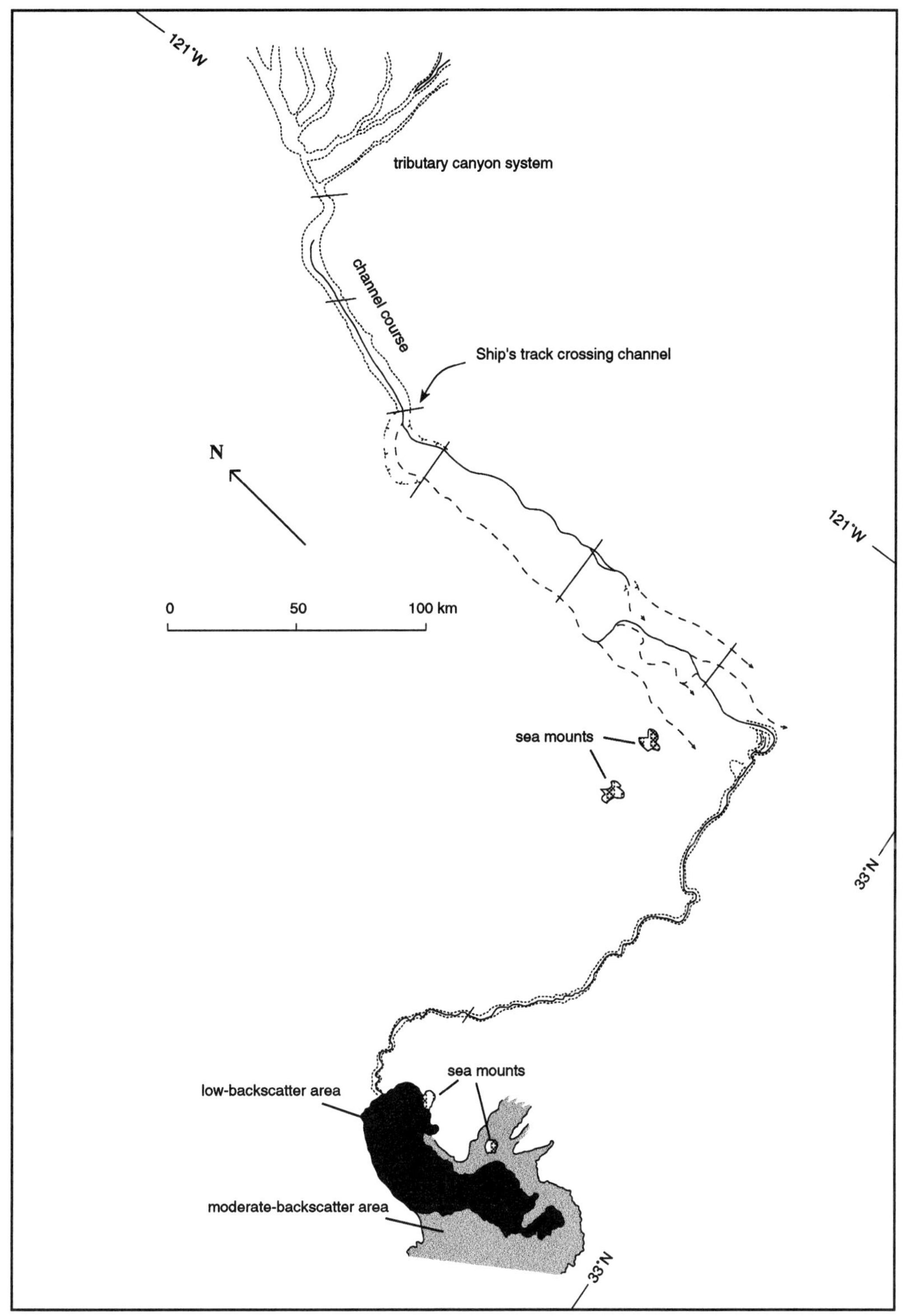
121°W
tributary canyon system
channel course
Ship's track crossing channel
N
0
50
100 km
121°W
sea mounts
33°N
sea mounts
low-backscatter area
moderate-backscatter area
33°N

Bering Sea channels

Age:	Modern
Fan position:	Canyon to fan
Widths (km):	2.1 - 6.2
Depths (m):	5 - 350 m
Length:	400 km
Aspect Ratios:	10 - 1000
Max. sinuosity:	1.22

The Bering Sea Basin, located between the Aleutian volcanic islands and the Bering shelf, contains three major distributary and depositional lobe systems, as well as many minor canyons and gullies (Bering Sea EEZ-SCAN Scientific Staff 1991, Kenyon & Millington 1995). The main feeder channel systems are the Bering Channel, the Umnak Channel and the Pochnoi Channel, which flow westwards and northwards to the deepest part of the basin near the Bowers Ridge. (Kenyon & Millington 1995). Contrasting styles of depositional lobe development between the Pochnoi and Umnak channels have been described by Kenyon and Millington (1995). The Pochnoi Channel ends in a depositional braided network of channels on the attached lobe, while the Umnak Channel has a "zone of erosion" between the channel mouth area and its depositional lobe. The erosion zone contains large crescentic scours, up to 2 km wide and 5 m deep, and linear bedforms (parallel to flow direction) believed to have formed from erosive flows.

The channels have a low sinuosity and form on the steep slopes of the northern margin of the Aleutian Island chain. The turbidite systems have a low volume sediment input from the relatively small drainage basin areas of the volcanic islands (Kenyon & Millington 1995). To the north, the Bering Shelf Slope also has many large canyon systems, but these are presently inactive (Kenyon & Millington 1995). The Umnak Channel contains areas of high backscatter associated with channel bends and confluences, and their shape suggests that they are depositional bedforms, likely to consist of coarse-grained material (Kenyon & Millington 1995) (Figure 5.7).

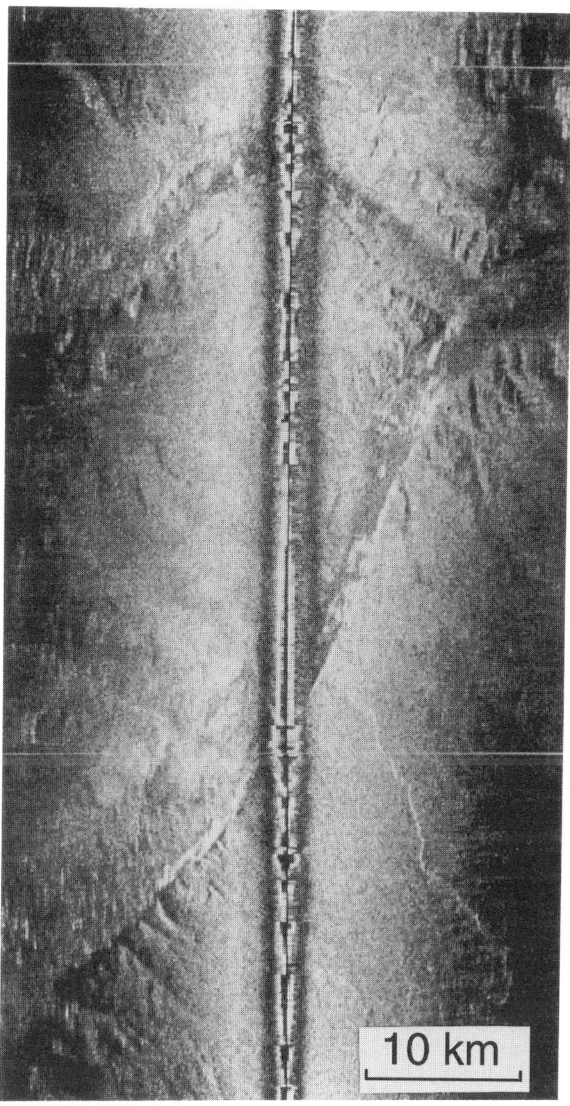

Figure 5.7. GLORIA sonograph and interpretation of a portion of the Umnak Channel, Bering Sea (Kenyon & Millington 1995). The sonograph shows the location of high-backscatter elements within the channel at locations including the channel confluence and channel bends. One interpretation is that these areas of high-backscatter are intra-channel bars. Flow direction is towards the north and west (towards the right in the sonograph).

Mississippi Channel, Gulf of Mexico

Age:	Modern
Fan position:	Canyon to fan
Widths (km):	3.8, 4.1, 4.5, 2.1, 1.1, 1.7, 1.2, 1.2
Depths (m):	20, 17, 27, 25, 20, 6, 50, 12
Length:	600 km
Aspect Ratios:	188, 243, 167, 83, 56, 281, 24, 100
Max. sinuosity:	1.74

The Mississippi Channel is the present day channel traversing the Mississippi Fan. The Mississippi Fan is located in the Gulf of Mexico, and fed from the extensive Mississippi drainage basin and delta. The Mississippi Fan covers an area 300 000 km^2 (Moore *et al.* 1978), and has been the focus for much research over the last two decades, with most notably the Deep Sea Drilling Project Leg 96 survey undertaken in 1983, and the EEZ-SCAN 85 GLORIA survey of the entire Gulf of Mexico basin undertaken in 1985. The fan is surrounded by the Florida Escarpment to the east, the Sigsbee Abyssal Plain to the west, and the Florida Abyssal Plain and Campeche Escarpment to the south. The fan sediments are late Pliocene and Pleistocene in age (Stuart & Caughey 1976, Feeley 1984, Shaub *et al.* 1984, Bouma *et al.* 1985b, Pickering *et al.* 1986b, Stelting *et al.* 1986).

The Mississippi Channel has been described by Garrison *et al.* (1982), Prior *et al.* (1983), Kastens and Shor (1985), O'Connell & Normark (1986), Pickering *et al.* (1986b), O'Connell *et al.* (1991), and Twichell *et al.* (1991). The present day leveed Mississippi Channel develops from the mouth of the Mississippi Canyon. It is 600 km long, and ranges in width from 0.5 to 4 km, and 10 to 50 m deep. GLORIA backscatter strengths from the channel floor are variable (Twichell *et al.* 1991). On the upper and middle fan, the channel is partially buried by an extensive slide feature, showing moderate- to high-backscatter (Figure 5.8). The slide was first discovered by Walker and Massingill (1970), and has since been mapped in more detail, showing it to be up to 95 m thick (Bryant *et al.* 1985), and even more extensive than was originally thought (Twichell *et al.* 1991). At the distal end of the channel, the channel

broadens, and a narrower (0.3 km) and straighter channel can be traced a further 50 km to a distributary system of minor channels with splay- and finger-shaped distributary ends (O'Connell *et al.* 1991, Twichell *et al.* 1991).

A sequence of nine depositional units have been established from relationships observed on 3.5 kHz and GLORIA records (Twichell *et al.* 1991). The youngest is the Walker-Massingill slide unit (mentioned above), with eight older lobes which show a general upfan trend in deposition succession (Twichell *et al.* 1991) (Figure 5.9). Depositional lobe 6 has particularly well preserved surface features seen on GLORIA sidescan sonographs, showing it to consist of an intricate network of small-scale bifurcating channels (Twichell *et al.* 1991) (Figure 5.10). The lobe appears to be deposited following a breach of main channel at a sharp bend on the middle fan area (Twichell *et al.* 1991).

A SeaMARC I survey of the distal part of this lobe shows a hierarchy of small-scale distributary channels covering its surface. The dimensions of the most distal channel are 10-20 m wide and <1 m deep (unresolvable on 3.5 kHz profiles). SeaMARC I and GLORIA data clearly shows that depositional lobes of the Mississippi Fan are deposited by a series of progressively increasing bifurcating, and rapidly switching channel distributary system, rather than deposition from sheet flow processes as was originally predicted (Twichell *et al.* 1991).

During the Deep Sea Drilling Project Leg 96, nine sites on The Mississippi Fan were drilled. These consisted of two middle fan channel sites (621 and 622), overbank sites (617 and 620), lower Mississippi Fan channel sites (623

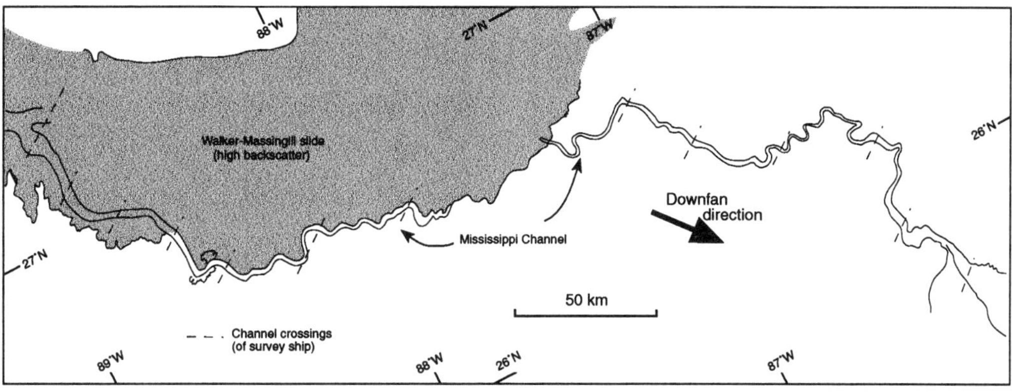

Figure 5.8. The course of the present day Mississippi Channel has been mapped using GLORIA sidescan sonar. The channel is partially buried by the recent extensive slide deposit, the Walker-Massingill slide. The Mississippi Canyon feeds the leveed fan valley from the left hand side and the downfan direction is to the right. Channel sinuosity gradually increases down-channel to a maximum, and then decreases. This trend is due to the effects of gradually decreasing downfan gradients. At the channel termination area, the channel broadens and shallows, and a thalweg channel (0.3 km wide) is seen to continue another 50 km in a less sinuous course (Twichell *et al.* 1991). Feathery high backscatter splays can be seen at channel bends in the upper part of the channel. These may be part of the Walker-Massingill slide deposit, or formed from channel overbank processes.

and 624), channel termination sites on the lower Fan (614 and 615), and a drilling site on youngest "fanlobe" (616) (Bouma *et al.* 1986, Pickering *et al.* 1986b). Channel lag gravel deposits were retrieved from the base of cores 622 and 621 (about 200 m depth from seafloor) taken from the middle fan channels sites (Bouma *et al.* 1986) (see selected core photographs in Figure 4.2). Seismic reflection profiles of this area suggest that these are channel lag deposits form a sequence of vertically stacked channels, showing an overall fining-up sequences (Pickering *et al.* 1986b) (Figure 5.11).

The Plio-Pleistocene stack of Mississippi Fan sediments comprises 17 discrete sequences, equivalent to the "turbidite systems" of Mutti and Normark (1987). These have been identified from seismic reflection profiles (Weimer & Buffler 1988, Weimer 1991). The closely spaced network of seismic survey lines has allowed individual channel distributary systems from these sequences to be mapped out (see Figure 1.7)

Quantitative measurements of the aggradation, lateral migration, sinuosity and bifurcation parameters, show highly variable characteristics of the channel distributary system throughout depositional history of the fan (Weimer & Buffler 1988, Weimer 1991).

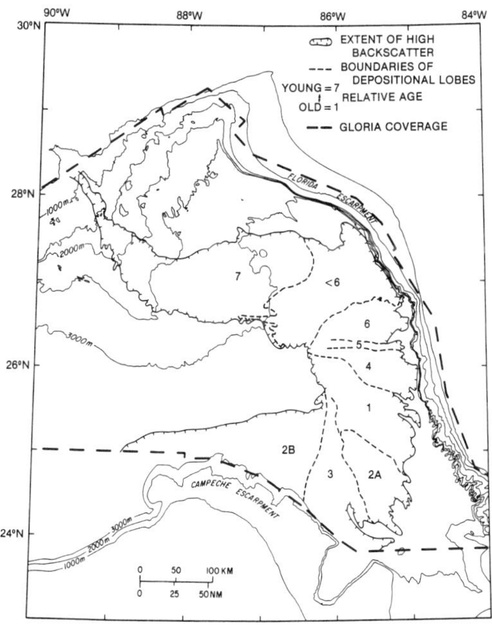

Figure 5.9. Map of the Mississippi Fan showing the present day distributary channel and the location of 9 depositional lobes (Twichell *et al.* 1991).

There are a series of general trends that can be identified from the sequence of channels, and these are believed to be most strongly controlled by Plio-Pleistocene eustacy (Bouma *et al.* 1984, Feeley 1984, Weimer & Buffler 1988, Weimer 1991). During periods of low sea-level, canyon backcutting into the slope results in increased sediment supply, and the formation of channel-levee systems (Steffens 1986). When sea-level rises, the backstepping of the deltas across the shelf, disconnects the direct sediment supply routes, resulting in the gradual infilling of the channel-levee systems. Decreasing sediment grainsize throughout the

development of individual channel-levee systems is inferred by Weimer (1991) based on the decreasing strength of the levee reflection character towards the top of each sequence. The decreasing volume and grainsize of flows during channel-levee evolution may have caused changes in channel sinuosity and levee relief (Weimer 1991). Indeed, Pickering *et al.* (1986b) explained the overall fining-upward sequence developed in the Mississippi Channel at DSDP Sites 621 and 622 as due to the glacio-eustatic rise in sea level from the Last Glacial Maximum into the Holocene.

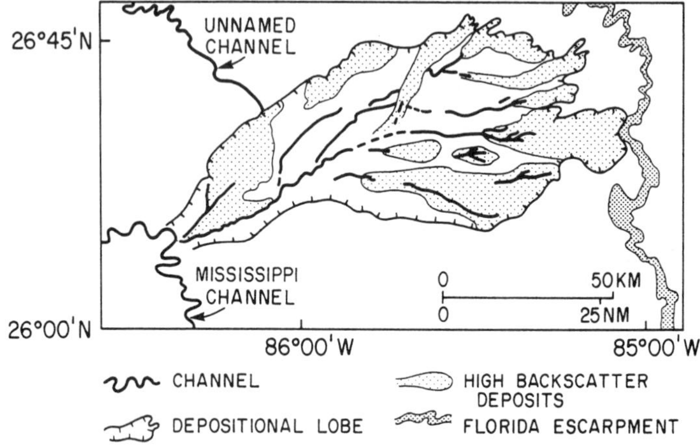

Figure 5.10. GLORIA interpretation of depositional lobe No. 6, Mississippi Fan (Twichell *et al.* 1991).

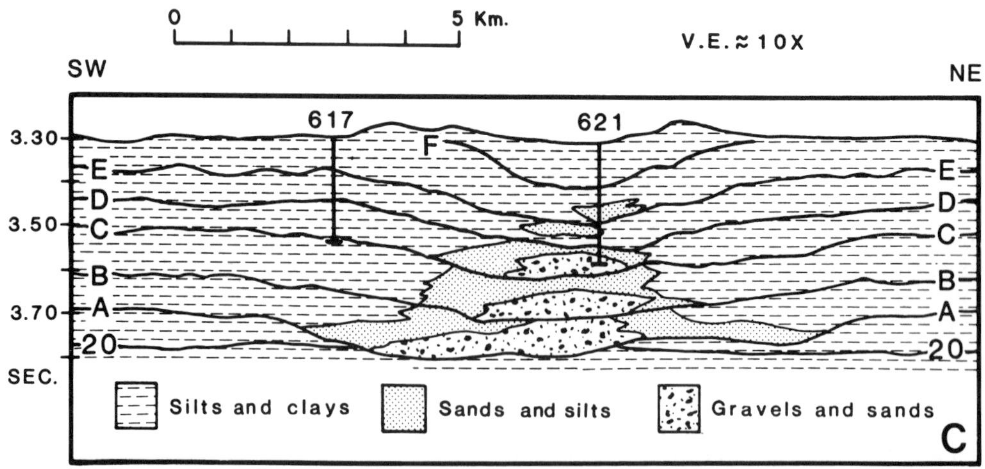

Figure 5.11. Model for channel-fill sequences of the Mississippi Channel inferred from seismic and core data. The channel shows a fining up sequence between successive seismic unit boundaries (after Stelting *et al.* (1985b).

DeSoto Channel, Gulf of Mexico

Age:	Modern
Fan position:	Canyon to fan
Widths (km):	0.6, 0.5, 0.4
Depths (m):	17, 15, 12
Length:	200 km
Aspect Ratios:	35, 33, 33
Max. sinuosity:	2.12

The DeSoto Channel is a highly sinuous channel, running approximately along the base of the Florida Escarpment (Figure 5.12). Very little is known about the origin and development of this channel. The leveed channel is connected to the DeSoto Canyon, but can not be traced to a source region on the shelf as it is buried by younger high backscatter material in the DeSoto Canyon area. It has been suggested, however, that the DeSoto Channel originates in the Mobile Bay, Alabama area (Twichell *et al.* 1991). The channel can be traced on GLORIA sonographs for 200 km, showing high backscatter from the channel floor (Twichell *et al.* 1991). To the south the channel is buried by the younger depositional lobe deposits of the Mississippi Fan.

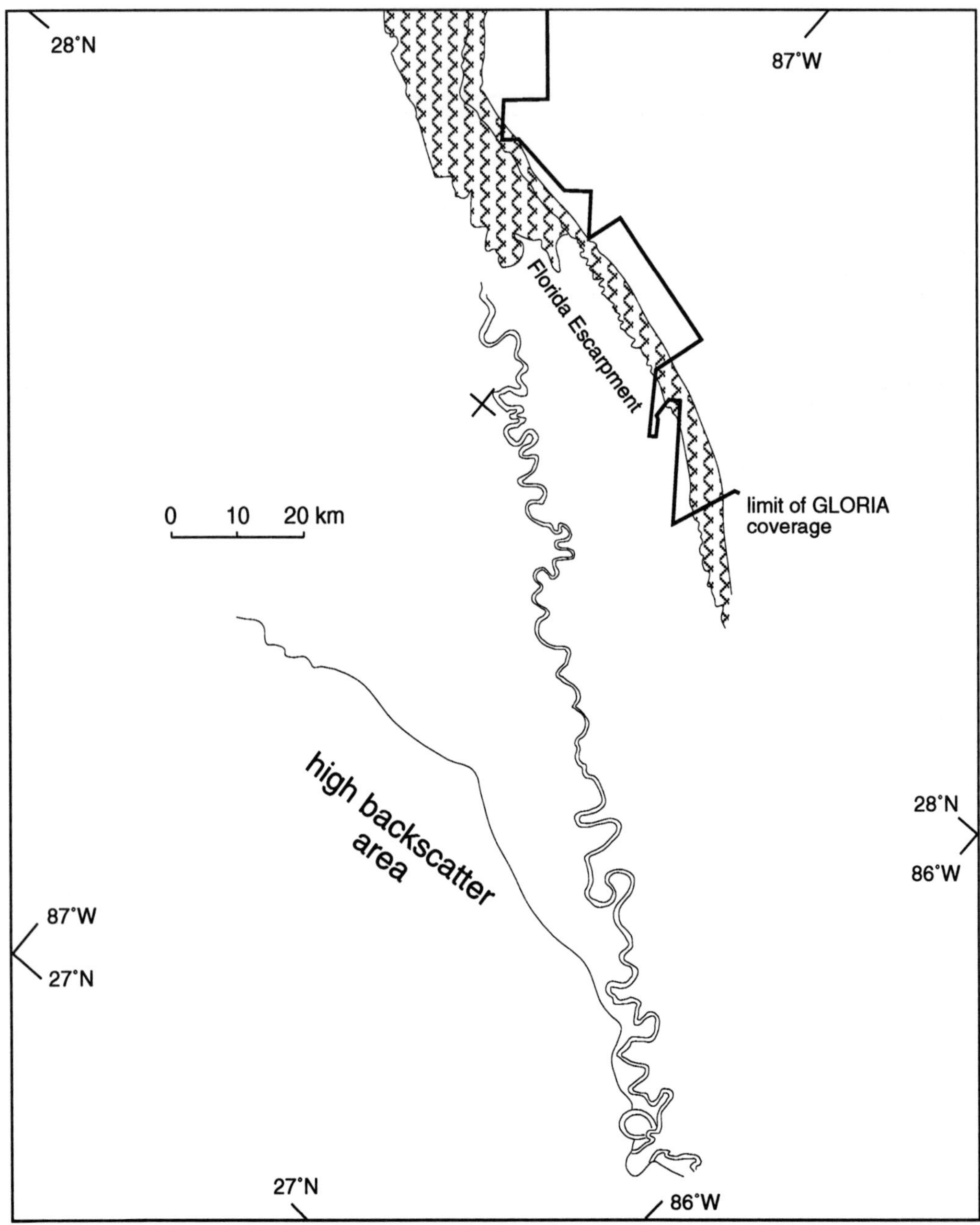

Figure 5.12. GLORIA interpretation showing the highly sinuous course of the DeSoto Channel, the Florida Escarpment and an area of high backscatter to the south.

Indus Fan channels, Indian Ocean

Age:	Modern
Fan position:	Canyon to fan
Widths (km):	1.5 (feeder channel "A"); 0.5-1 (channels A1, A2, A3)
Depths (m):	195 (feeder channel "A"); 7-100 (channels A1, A2, A3)
Length:	1000+ km
Aspect Ratios:	7.7 (feeder channel "A"); 8-50 (channels A1, A2, A3)
Max. sinuosity:	3.13

The Indus Fan is located in the Arabian Sea, north-west Indian Ocean, and covers an area of 1.1 x 106 km^2 (Kolla & Coumes 1985). The fan extends southwards from the India-Pakistan continental margin, bounded by the Chagos-Laccadive Ridge to the east, the Owen and Murray ridges to the west and north, and the Carlsberg ridge to the south. The fan is sourced by sediment from the Indus River system which drains the Himalayan mountains in a subtropical to temperate climate. The river extends a distance of 12,000 km from source to delta, resulting in much of its sediment being fine-grained and well sorted (Kolla & Coumes 1987). Present-day sediment discharge from the Indus River has been calculated at 460 million tonnes a^{-1} (Lisitzin 1972). Deposition on the fan commenced in Oligocene-Miocene time following the major uplift of the Himalayas (Kolla & Coumes 1985). The upper, middle and lower fan are all channelised, with channel size decreasing downfan. The profile of the fan shows an irregular upper fan area, cut by canyons and channels, and an upward bulge in the middle fan area (Figure 5.13) (Kolla & Coumes 1987).

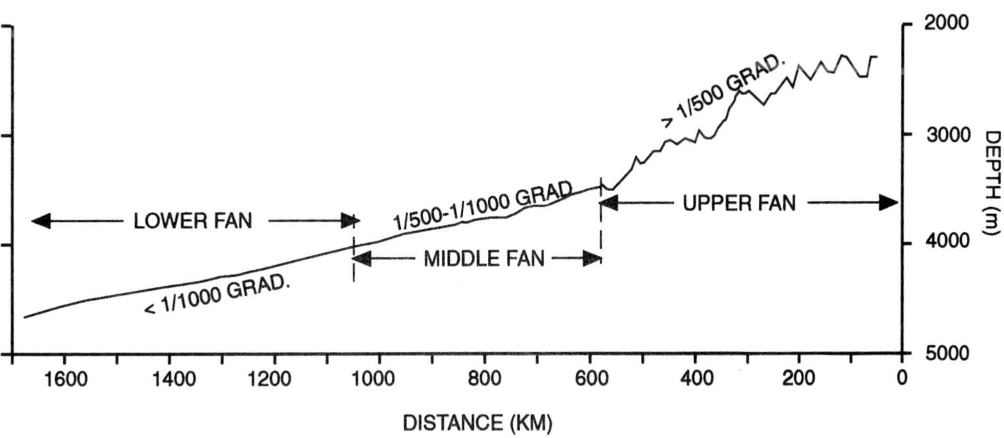

Figure 5.13. Profile of the Indus Fan showing the gradients of the upper, middle and lower fan divisions (redrawn after Kolla & Coumes 1987).

The Indus Canyon is erosional on the upper continental slope and shelf regions, while channels on the lower slope and upper fan are depositional (Kolla & Coumes 1985, McHargue & Webb 1986). The present canyon is believed to have formed during a period of low eustatic sea level, in the Pleistocene, and evolved from retrogressive sliding and slumping during the subsequent sea level rise (Kolla & Coumes 1987). At least two other (older) canyons exist on the continental slope, and have acted as major feeder systems to the fan (Kolla & Coumes 1985, 1987, McHargue 1991, Kenyon *et al.* 1995a). Upper fan channels have high levees (up to 100 m), and can be up to 8 km wide. Middle fan channels have 20-m-high levees, and lower fan channels have levees 8-20 m high (Kolla & Coumes 1987). GLORIA sonographs show that the upper fan channels are extremely sinuous. A hierarchy of canyon/channel feeder systems on the fan has been established by Kenyon *et al.* (1995a) (Figure 5.14). As observed from other fans (e.g. the Mississippi and Amazon fans), only one channel system is believed to be active at a time.

The present-day Indus Canyon feeds a series of channel levee systems whose relative ages have been established using 3.5 kHz data and by measuring the thickness of hemipelagic drape covering the channel-levee complexes (see Figure 5.14). The youngest is the "A" channel system (Kenyon *et al.* 1995a), and the older "B" channel system is believed to have been supplied from the ancient Saraswati River and Canyon (Kolla & Coumes 1987, Kenyon *et al.* 1995a). Channel system "A" is covered by an average of 0.4 m of mudstone drape, while channel system "B" and "C" are covered by ~4 m and ~12 m, respectively (Kenyon *et al.* 1995a). GLORIA sonographs show a series of smaller-scale channel-levee complexes form a radiating pattern, from a common site of avulsion at the mouth of Channel "A" (Kenyon *et al.* 1995a) (see Figure 5.14).

All the channel-levee complexes have higher backscatter channel floors compared with the surrounding overbank areas. Moderately high backscatter can be seen on the levee areas of the small-scale channel-levee complexes, and flat areas of high backscatter are observed in interchannel areas (Kenyon *et al.* 1995a).

There is a crude correlation between the location of coarse-grained material retrieved from cores and areas of high backscatter, suggesting that parts of the interchannel and levee bases may contain sandy deposits (Kenyon *et al.* 1995a). Figure 5.15 shows a GLORIA sonograph of the highly sinuous (sinuosity value of 3.1) upper Indus channel "A", with bright spots located at the channel bends. These high backscatter areas may be caused by the deposition of coarser-grained material, due to flow deceleration processes and, or, point bar deposition at channel bends.

The youngest of the channel-levee complexes (A1) has a depositional lobe, which shows an area of high and streaky backscatter, interpreted as a channel-mouth sandy lobe covered with a distributary network of low sinuosity channels. Channels in this area have been identified from 3.5 kHz records (up to 8 m deep, Kenyon *et al.* 1995a), suggesting that 30% of the area of this lower fan region is covered by channels (Kolla & Coumes 1987). The proportion of channels may be much higher as many of the streaky areas of high backscatter seen on the sonographs have no resolvable topography on the seismic records.

Channel avulsion occurs both progressively in one direction and also by jumps, and may result from a variety of processes, including sea level changes, tectonic uplift events in the Himalayas and Murray Ridge, Coriolis force, channel plugging from deposition of channel-wall slumping and debris flows, as well as autocyclic aggradation and channel meander processes (Kolla & Coumes 1987). Kenyon *et al.* (1995a), suggested that avulsion occurs from natural autocyclic processes for the small-scale channel-levee complexes, whereas avulsion in the larger channel-levee complexes results from sea-level changes.

Seismic profiles of the upper and middle fan channel-complexes show wedge shaped reflection packages, with high amplitude random reflections overlain by reflection free zones. Together with the sand and silt recovered in cores from the channels, this suggests that the channels have coarse-grained deposits at their bases which fine upwards (Kolla & Coumes 1987). This trend has also been observed from multi-channel seismic data of a series of older channels, stemming

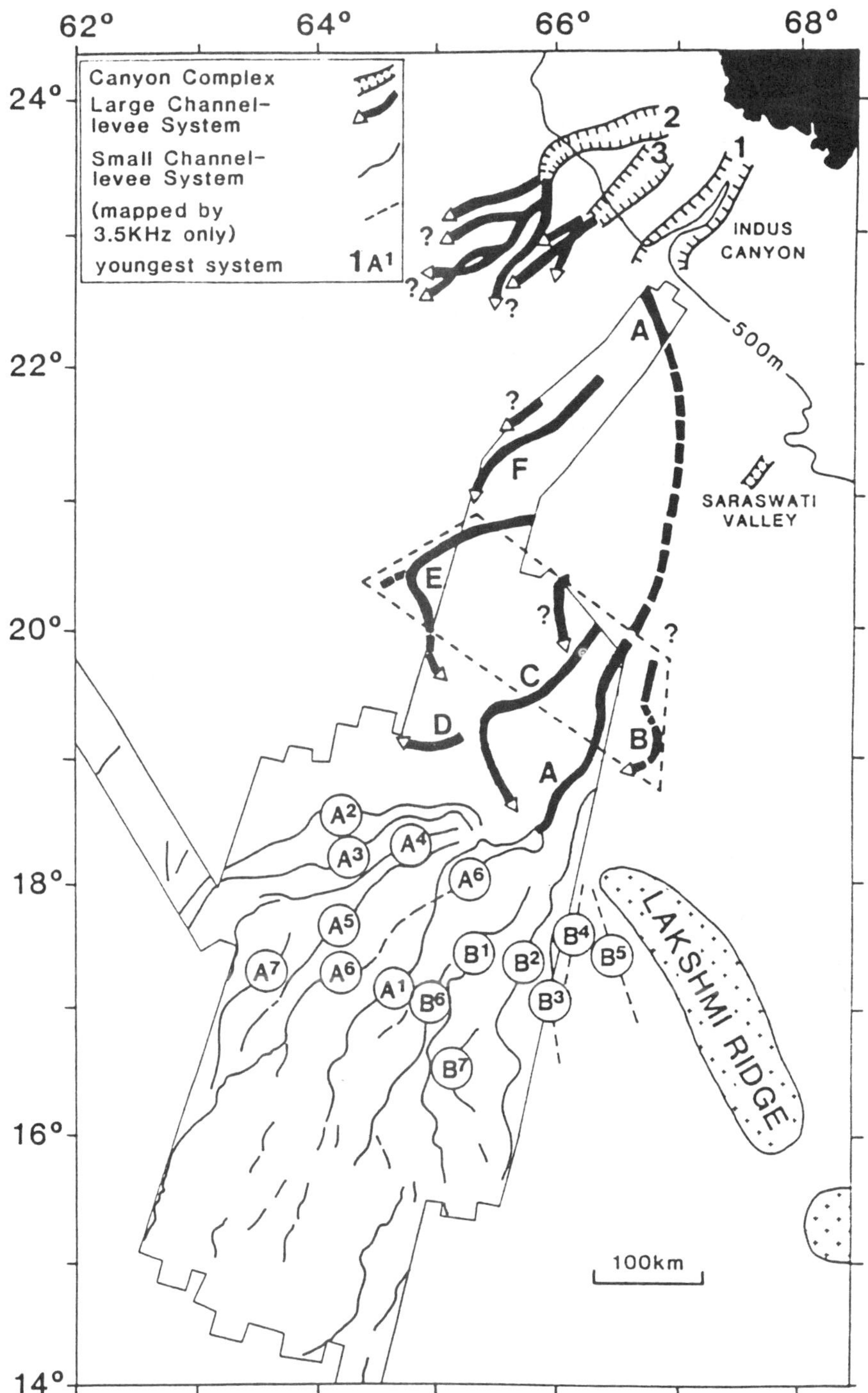

Figure 5.14. Map of the Indus fan channels and canyons, showing the hierarchy of distributary systems. Canyon 1 (youngest) supplies channel system A, which branches into a network of distributary channels (A1-A7) (Kenyon *et al.* 1995a). Channel system B was probably linked to the Saraswati valley.

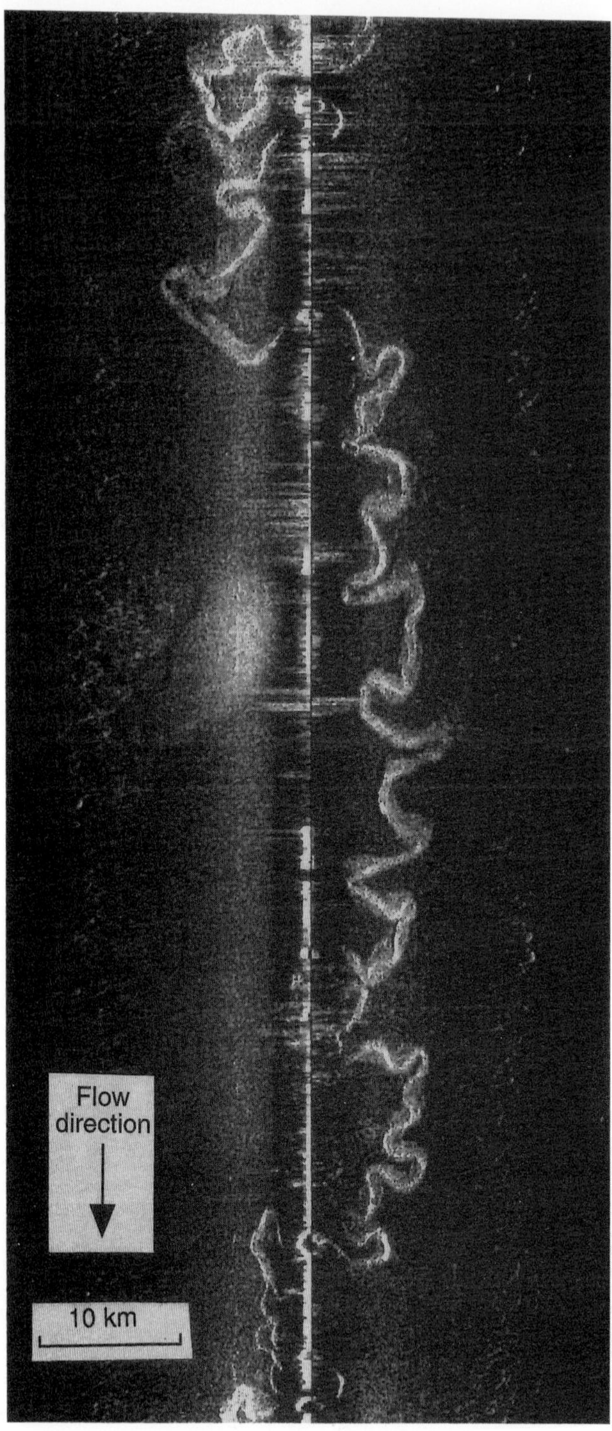

Figure 5.15. GLORIA sonograph of the upper Indus fan channel ("A"), showing the highly sinuous channel course and the location of high-backscatter elements at channel bends (Kenyon *et al.* 1995a). Note the asymmetry of the meander bends (up-valley skew), a property commonly observed in fluvial meanders.

from the mouth of Canyon 2 (see Figure 5.14 for location). McHargue (1991) has demonstrated that the history of these channels involves an overall fining-up due to waning flow strength (Figure 5.16). During lower sea-level stands, channels develop was more active due to the increase in sediment supply, and the finning-upward sequences result from decreasing volume and size of turbidity currents during sea-level rise (Kolla & Coumes 1987, McHargue 1991). The increase in channel branching, and associated decrease in sinuosity, on the depositional lobe may be due to less fine-grained sediment in suspension in the turbidity currents (Kolla & Coumes 1987). (Figure 5.17).

Sand deposition on the Indus Fan, therefore, can be expected in channel lag deposits (interpreted from seismic facies, and cores), point bars related to channel bends, interchannel areas (from their association with high backscatter areas), sandy lobes containing minor distributary channels, and at the base of levee sequences (interpreted from high amplitude reflectors on seismic records of levees) (Kenyon *et al.* 1995a) (Figure 5.18).

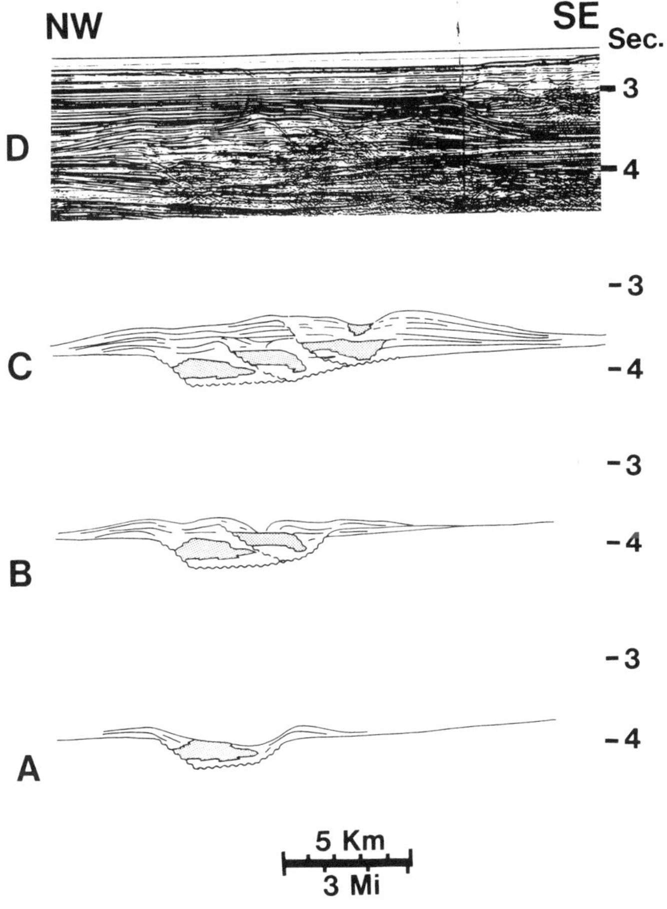

Figure 5.16. Seismic facies showing the evolution of a buried channel fed by Canyon 2 (McHargue 1991). (a) The channel initially had an erosional base and small levees, which then developed and grew by south-eastward lateral accretion (b). Lateral accretion continued with increased channel aggradation, and the formation of extensive levees (c). High amplitude reflections (dot pattern), implies relatively sand-rich facies are located in channel axis, and that an overall fining-up sequence developed. (d) Uninterpreted seismic profile.

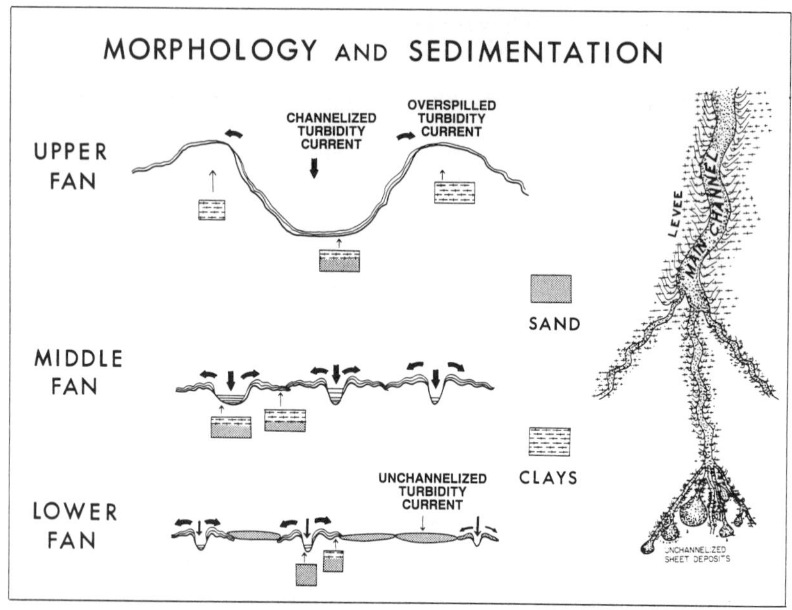

Figure 5.17. Morphology and sedimentation in the upper, middle and lower fan channels of the Indus Fan (Kolla & Coumes 1987).

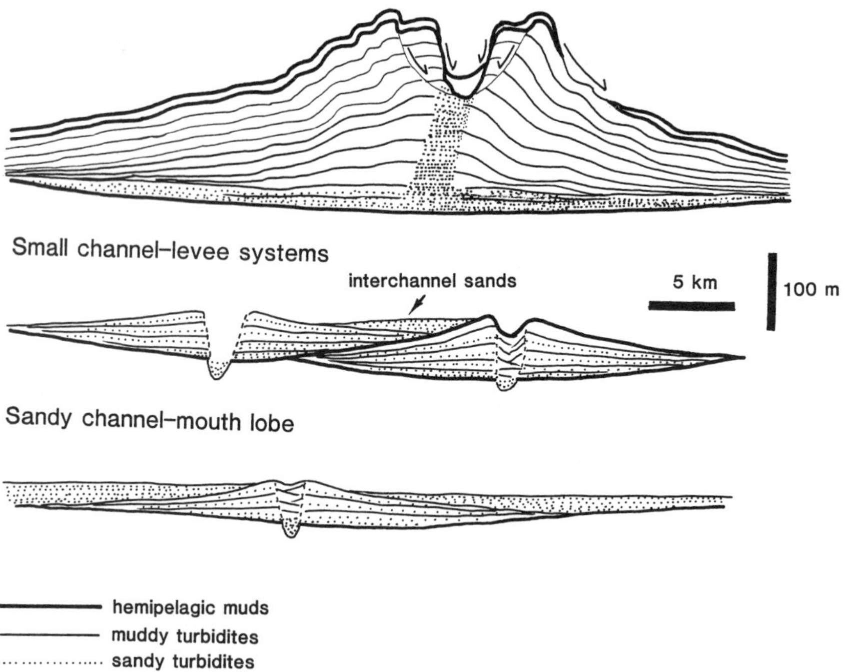

Figure 5.18. Profile models for the large channel-levee systems, small channel-levee systems and sandy channel-mouth lobes. All channels show an aggradational character and a vertical stacking of channel axis deposits (Kenyon *et al.* 1995a).

Amazon Fan channels, Atlantic Ocean

Age:	Modern
Fan position:	Canyon to fan
Widths (km):	0.4 - 2.5
Depths (m):	2 -200
Length:	300+ km
Aspect Ratios:	10 (average)
Max. sinuosity:	2.6

The Amazon Fan covers an area of 330,000 km^2, and is similar in size to the Mississippi Fan. It has been the subject of many studies, principally by Damuth and co-workers (see Damuth & Kumar 1975a, Damuth & Embley 1981, Damuth *et al.* 1983a, 1983b, Damuth & Flood 1984, Damuth *et al.* 1988, Manley & Flood 1988, Flood *et al.* 1991). Deposition commenced in the Miocene (Damuth & Kumar 1975a). The fan is elongate and covered by large sediment slides. The upper fan contains the Amazon Submarine Canyon feeding the Central Channel. The large leveed Central Channel feeds two major middle fan channel complexes, the Western Channel Complex and the Eastern Channel Complex (Figure 5.19). The middle fan channels are highly sinuous (maximum sinuosity 2.5), and show a complex network of bifurcating and avulsed channels.

Only one channel-levee system is active at any time (Damuth *et al.* 1983b), and probably only during periods of relatively low sea level. The Amazon Canyon is presently inactive due to the high sea level stand. During periods of low sea level, the Amazon River is able to extend across the shelf and feed directly into the Canyon. At such times, quasi-continuous turbidity currents, generated from the direct fluvial input to the fan, may have been the main cause for the evolution of the highly sinuous channel systems (Damuth & Kumar 1975b). Thus sea-level variations can be demonstrated as being responsible for the major channel bifurcations on the upper fan, although many of the bifurcations of smaller channels and the deposition of major slides are apparently unrelated to sea level changes (Flood *et al.* 1991) (as also observed for the Indus fan channels). Channel avulsion on the middle fan occurs by breaching the channel levee, commonly on a meander bend, and the channel finding a topographically lower course (Figure 5.20). In this way channel-levee deposits build up by overlapping and coalescing (Damuth *et al.* 1983b).

Numerous small-scale sinuous channels deposit sand-grade sediment on the lower fan region, while deposition from overbank processes on the middle fan is essentially muddy (Damuth *et al.* 1988). The Amazon Fan, therefore, is an example of a high efficiency fan, *sensu* Mutti (1979).

Recent drilling on the Amazon Fan (ODP Leg 155) cored 17 sites (ODP Sites 930-946), with the principal initial conclusions being (Flood, Piper & Klaus *et al.* 1995), :

- Channel-levee-overbank systems on the Amazon have fast rates of development and aggradation. Initial results suggest that major shifts in channel position during the late Pleistocene, caused by avulsion processes on the upper fan, occurred approximately every 5-10 kyr.
- Glacial-age sediment accumulation rates associated with the levees of abandoned channels show relatively low values of 1-3 m kyr^{-1}, in contrast to rates of 10-25 m kyr^{-1} for active channels: sand-rich lobe deposits down-channel from the latter have average sediment accumulation rates of 2 m kyr^{-1}.
- The pelagic calcareous clays that mantle the Amazon Fan during high-stands in eustatic sea level have typical sediment accumulation rates of ca. 0.1 m kyr^{-1}.
- There appears to be no simple relationship between the occurrence and location of major (>100 m thick) debris-flow deposits and eustatic sea-level change.

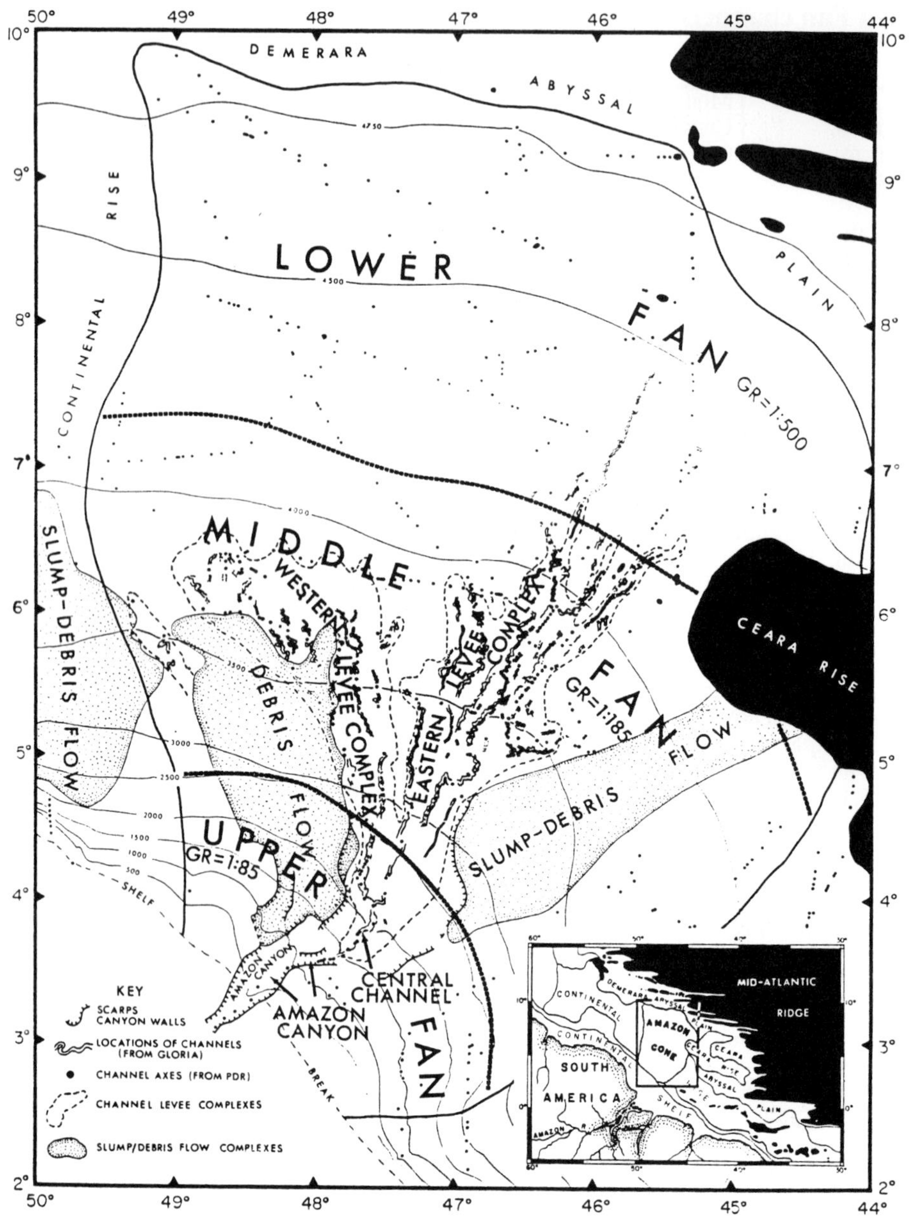

Figure 5.19. Map of the Western and Eastern channel-levee complexes of the Amazon Fan (Damuth *et al.* 1983a).

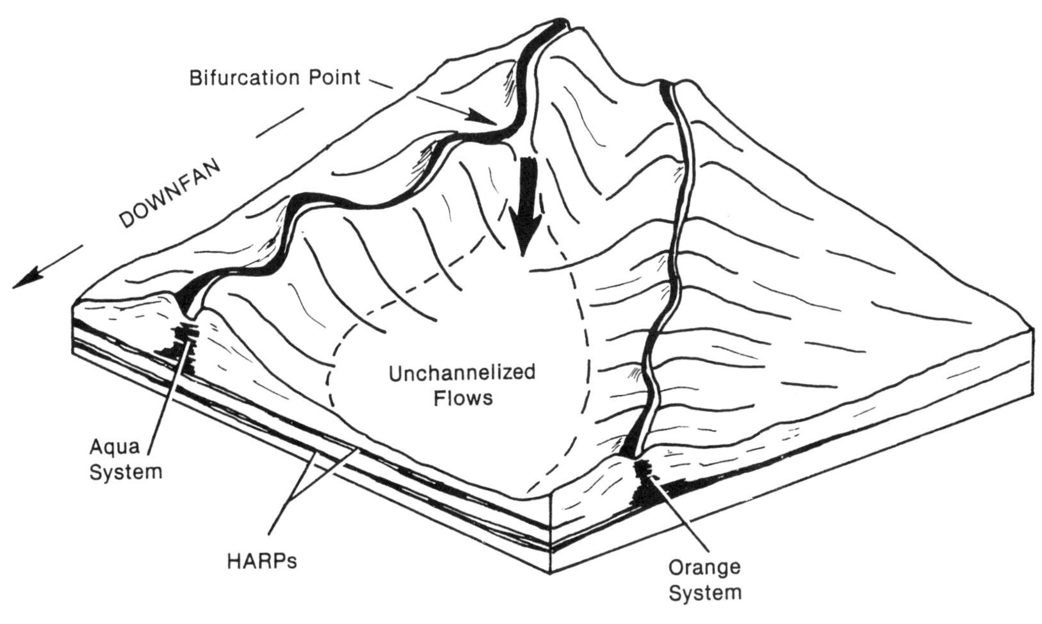

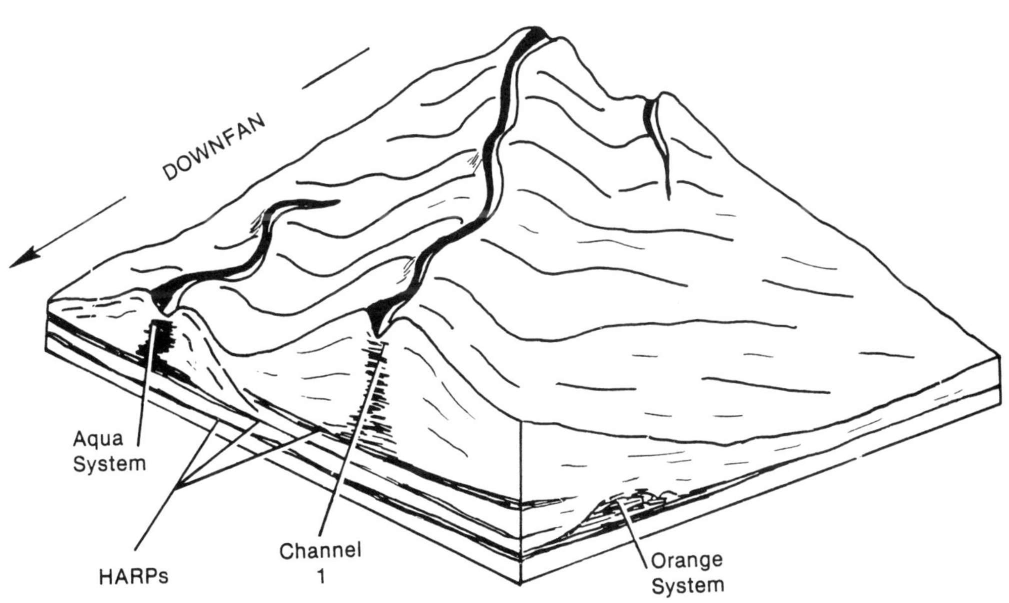

Figure 5.20. Model to show the processes of channel breaching, abandonment and the development and growth of a new aggradational channel (Flood *et al.* 1991). High Amplitude Reflection Planes (HARPs) have been identified and channel nomenclature is from (Flood *et al.* 1991).

Baltimore-Wilmington, Hudson and Atlantis channels, East Coast USA

Age:	Modern
Fan position:	Canyon to fan
Widths (km):	0.75 - 4
Depths (m):	15 - 500
Length:	250+ km
Aspect Ratios:	8 - 40
Max. sinuosity:	1.48

Many canyons are incised into the continental slope off the eastern coast of the United States. Vetch and Smith (1939) produced a contour map of the East Coast Continental Slope, showing dendritic drainage patterns of submarine canyons. Further studies have resulted in the discovery of even more canyons and gullies, e.g. GLORIA surveys (Twichell *et al.* 1980, EEZ-SCAN 87 Scientific Staff 1990), detailed bathymetry surveys (McGregor *et al.* 1982, Pratson *et al.* 1994), extensive mapping of the slope and rise using 3.5 kHz seismic reflection character (Pratson & Laine 1989), and the SeaMARC I survey of the Wilmington Canyon (McGregor *et al.* 1982).

The main canyons off the eastern U.S. continental margin are the Hatteras Canyon system, the Norfolk-Washington Canyon system, the Baltimore-Wilmington Canyon system, and the Hudson and Atlantis canyons (Figure 5.21). The sediments for these canyons are derived from rivers that drain the Appalachian mountains and carry sediment out onto the shelf. The larger canyons extend back into the shelf, whereas the numerous minor canyons and gullies originate from headwall slumping on the continental slope (Pratson & Laine 1989).

The Baltimore-Wilmington Canyon system shows a dendritic drainage pattern of minor canyons and gullies on the upper continental slope. These canyons feed the slightly sinuous Baltimore-Wilmington channel. The smaller canyons, in the vicinity of the Baltimore-Wilmington Canyon, such as the South Wilmington and North Heyes canyons, are not as sinuous despite having similar slope gradients. These smaller canyons, which only extend back onto the slope area, are believed to be morphologically less mature than the older Baltimore-Wilmington Canyon System (McGregor *et al.* 1982). The Hudson and Atlantis canyons also have developed channel systems extending onto the lower continental rise. GLORIA sonographs of these channels show that channel wall failure, as sediment slides and slumps, has been active in controlling their morphology.

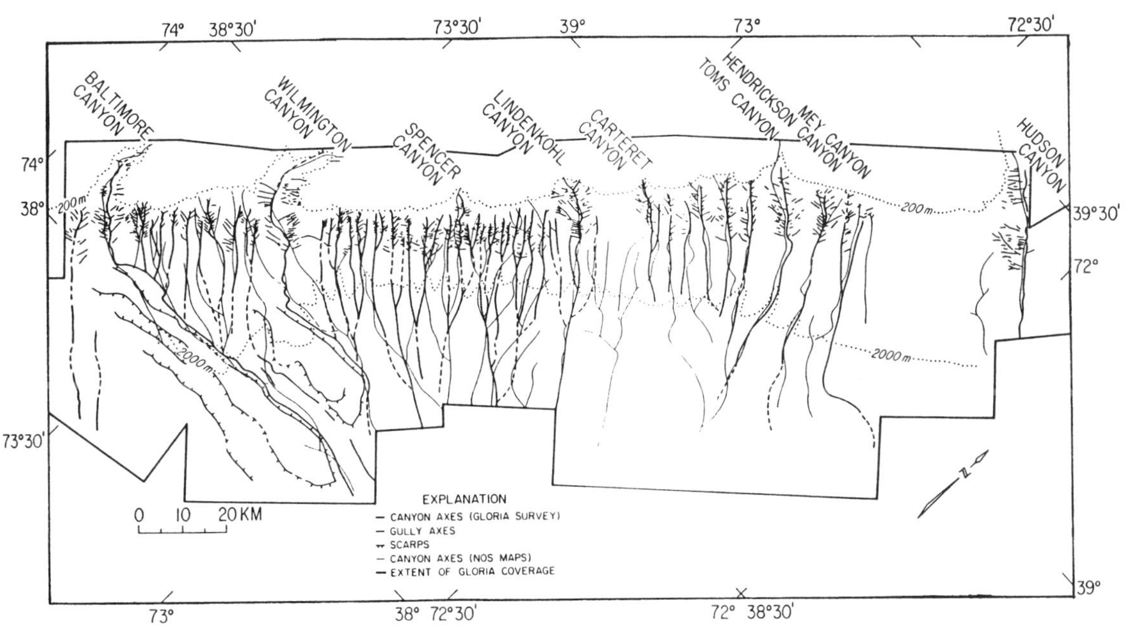

Figure 5.21. GLORIA map of the canyons and slope deposits of the east coast USA between Baltimore and Hudson canyons (Twichell & Roberts 1982).

Laurentian Fan channels

The Laurentian Fan lies on the eastern Canadian continental slope, and receives sediment supplied from the continental slope and the St. Lawrence River, via the Laurentian Channel incised into the continental shelf between Nova Scotia and Newfoundland. The Laurentian Fan contains four principal channel-levee complexes, the Western, Central and Eastern fan valleys, and the Grand Banks Valley (Piper *et al.* 1985, Hughes Clarke *et al.* 1990). The fan also has numerous minor channel systems, including tributaries originating from nearby continental slopes (e.g. St. Pierre Valley), and sinuous valleys of the Intervalley Divide (Figure 5.22). The Intervalley divide lies between the Western and Eastern fan valleys, and contains a network of sinuous valleys, believed to be the result of mass-wasting processes (Piper *et al.* 1985). These features may represent a relatively mature morphology, as these processes are likely to act over considerable periods of time (Piper *et al.* 1985).

Below the upper slope channels of the Eastern Valley there is a large relatively flat field of gravel waves (Piper *et al.* 1985). The waves are typically 2-5 m high and 50-100 m in wavelength, and aligned parallel to contours. Small channels from the upper part of the Eastern Valley appear to have deposited sand ribbons on the gravel substrate, 100-500 m in width and up to 25 km in length (Piper *et al.* 1985). The gravel waves described by Piper *et al.* (1985), like the slide features on the upper slope of the Eastern Valley, appear to have no discernible pelagic drape (from either sidescan records or core samples), and seismic profiles in the area show a contrasting lack of such features in the upper 50-100 m of sediment. It is concluded, therefore, that the slide features probably resulted from the 1929 Grand Banks earthquake (its epicentre lying in the upper slope region near the Eastern Valley), and the gravel waves formed from subsequent flows depositing and reworking gravel in a traction carpet further downslope (Piper *et al.* 1985).

The Western, Central and Grand Banks valleys are typically 5 km wide with a V-shaped morphology, while the Eastern Valley has an average width of 25 km and a flat channel floor (Figure 5.23). The Eastern

Valley was the main conduit for the turbidity current resulting from the 1929 Grand Banks earthquake, and is described in detail below.

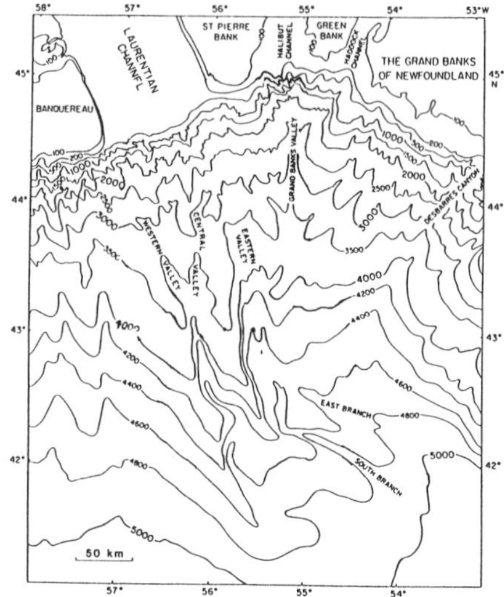

Figure 5.22. Bathymetry of the Laurentian Fan showing shelf and slope areas and fan valleys (Hughes Clarke *et al.* 1990).

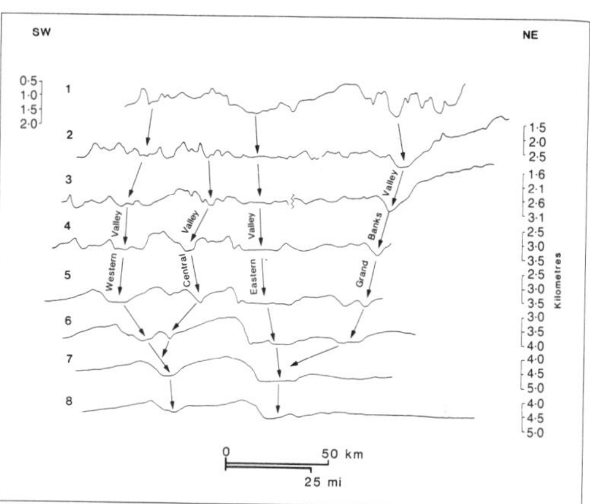

Figure 5.23. Profiles across the Laurentian fan between 1000 m and 4500 m water depth (Masson *et al.* 1985).

Eastern Valley of the Laurentian Fan

Age:	Modern
Fan position:	Canyon to fan
Widths (km):	25 (average)
Depths (m):	950 (maximum)
Length:	250+ km
Aspect Ratios:	50 (average)
Max. sinuosity:	1.30

The Eastern Valley of the Laurentian Fan is the largest of the three principal valley systems. Its has an average width is 25 km and maximum levee relief of 950 m (Hughes Clarke *et al.* 1990) (Figure 5.24). The upper Valley extends to the 2000 m contour (the up-slope edge of the gravel wave field). The head of the valley consists of a broad area of cuspate slide scars, below the shelf break, that emanate downslope-converging gullies. South of the 1500m contour the gullies evolve into narrow erosional channels, with increasing elevation of confining walls, to a flat-floored, abruptly margined, 20 km wide valley (Hughes Clarke *et al.* 1990). The middle valley floor extends to the point of valley bifurcation at the 4300 m contour. Seabeam swath bathymetry has identified numerous, discontinuous non-meandering channels preferentially located in the valley axis, and adjacent to the main valley margins, in some places eroding into adjacent abandoned levee material (Hughes Clarke *et al.* 1990). Piston cores of valley floor sediments show muddy sands and gravels (Hughes Clarke 1988). The lower valley splits into two valley branches, the East, and South Branch, and is separated by the overbank deposits of the Interbranch High (Hughes Clarke 1988). The two valleys contain a single flat-floored channel thalweg, and valley relief decreases down-fan, until they eventually merge into the proximal fan lobe at the 4900 m contour (termed the Valley Termination Zone by Piper, Stow & Normark 1984).

Valley floor channels

The discontinuous valley floor channels of the upper and middle fan valleys are not likely to have been the result of confined turbidity currents. Rather, these channels are erosional features, resulting from enhanced secondary localised flow circulation (Hughes Clarke *et al.* 1990). These channels show a down-channel increase in erosion (increase in channel relief), to a peak on the middle valley floor, and subsequently decrease towards the lower valley, where the only noticeable erosion, i.e. currently active, is that on the outside bends of the valley branches.

Gravel waves

The gravel waves occur in bedform fields covering 80% of the Eastern Valley floor (Piper *et al.* 1988). The gravel waves are asymmetrical dune bedforms, orientated transverse to flow, with straight to mildly sinuous elongate crests (Hughes Clarke *et al.* 1990). There is a general down-valley increase in average bedform wavelength, in the middle valley floor (30 m to 70 m), and they typically show amplitudes between 5 and 10 m (Hughes Clarke *et al.* 1990) (Figure 5.25). Photographs from the submersible dives show the waves to consist of boulder conglomerates, clast-supported pebble and cobble gravel, graded gravel beds and mud-clast rich pebble conglomerates, with no obvious cross-stratification structures (Hughes Clarke *et al.* 1990). These bedforms are comparable to pebble and cobble gravel deposits in the Var Canyon (Malinverno *et al.* 1988). The Var Canyon deposits are also attributed to a recent (1979) catastrophic slide event. Beyond the Lower Valley floor, the gravel waves disappear, and are replaced by macro-dune bedforms (mean wavelength of 300 m), consisting of moderately- to poorly-sorted granule gravel deposits (Hughes Clarke *et al.* 1990).

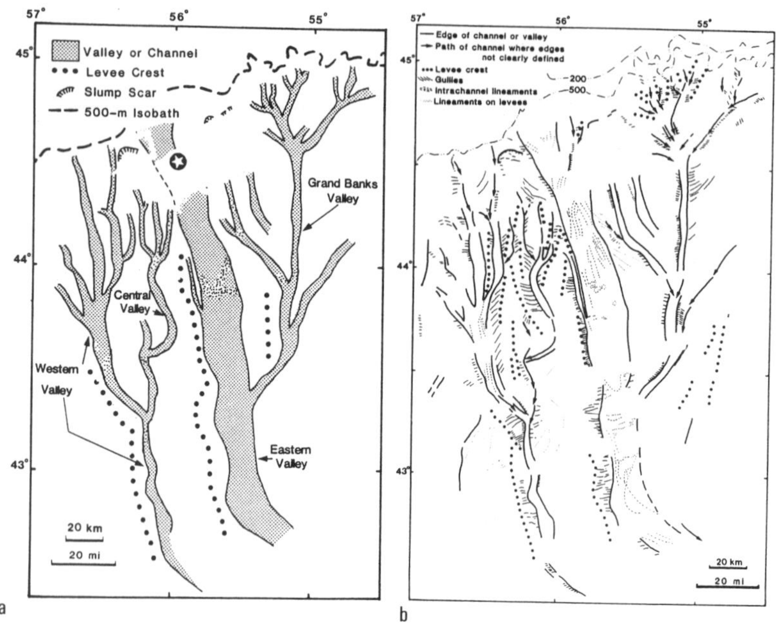

Figure 5.24. (a) Simplified and (b) detailed maps of the morphology of the Laurentian fan channels (Masson *et al.* 1985).

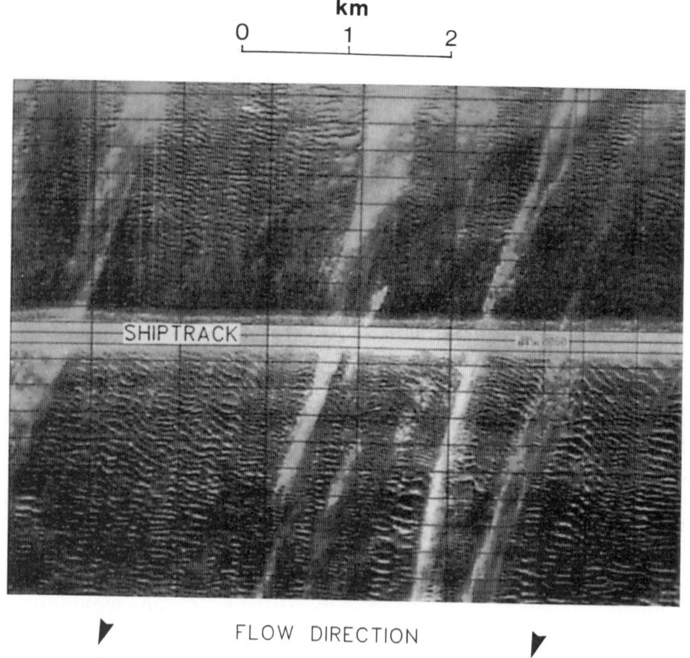

Figure 5.25. SeaMARC I swath image of the gravel-wave fields (dark) and sand ribbons (light) on the floor of the Eastern Valley (Hughes Clarke *et al.* 1990)

Flow processes

The gravel wave material is believed to pre-date the 1929 turbidity current, but the high-velocity flow of this sediment gravity flow event resulted in the formation of these features. The down valley change in gravel bedform wavelength, in the middle valley floor, suggests the flow was accelerating on this part of the fan. Since the gradient in this area decreases, the increase in flow acceleration can be thought of as a consequence of flow entrainment of sediment produced from erosion. As valley-floor width increases on the lower valley floor, turbidity currents became less confined, resulting in a loss of hydraulic head. Deceleration resulted in the erosional flows transforming to depositional flows, and a change in the resultant bedforms, from the gravel waves, to the macro-dunes on the depositional lobe.

Rhône Channel, Western Mediterranean

Age:	Modern
Fan position:	Canyon to fan
Widths (km):	0.5 - 1.2
Depths (m):	10 - 100
Length:	120 km
Aspect Ratios:	2.8 - 11
Max. sinuosity:	1.48

The Rhône Fan is located south of the Rhône River delta in the north-west Mediterranean Sea. Fan sedimentation commenced in the early Pliocene, with fan sediments lying directly above Messinian evaporates. Three major submarine canyons have been identified on the upper Rhône Fan, the Grand-Rhône, Arles and Petite-Rhône canyons (Droz & Bellaiche 1985). The Petite-Rhône Canyon has been the most active of all the canyons in this area in the recent past, although relatively inactive during the present high-stand in sea level. The Petite-Rhône Canyon and upper fan valley have been mapped using Seabeam bathymetry (Bellaiche *et al.* 1983, Droz & Bellaiche 1985, Bellaiche *et al.* 1986a). A GLORIA survey was carried out in 1983, and a SeaMARC high-resolution sidescan survey in 1984, which revealed the sinuous nature of the upper Rhône Canyon, submarine valley and thalweg (O'Connell *et al.* 1991) (Figure 5.26). The Rhône Channel thalweg has uncharacteristically arcuate and cuspate meander bends. Similarities have been made between the morphology of the Rhône Channel with that of Martian Channels rather than terrestrial channel forms (Bellaiche *et al.* 1986b).

Seismic profiles of the Rhône Channel show the aggradational build up of high levees, up to 500 m above the level of the surrounding seafloor (O'Connell *et al.* 1991). Sediment failure as slumps and other mass-wasting is common on the steep sides of the upper fan channel levees (O'Connell *et al.* 1991). At one point along the channel, avulsion has occurred in the recent past, resulting in abandonment of the channel-levee system downfan from the avulsion point, and the development of a new channel (Figure 46). The younger channel is relatively straight compared to the tight meanders of the upper Rhône Channel (Figure 47). The breaching of the levee has substantially lowered the depositional base level for the channel-levee system, and resulted in entrenchment of the Rhône Channel thalweg, by erosion. The uncharacteristically steep sides and cuspate or scalloped shaped meander curves of the present Rhône Channel thalweg are a result of the Rhône Channel currently in a process of re-establishing an equilibrium through active entrenchment of the fan valley (O'Connell *et al.* 1991). The Rhône Channel, therefore, is an example of a juvenile entrenched system.

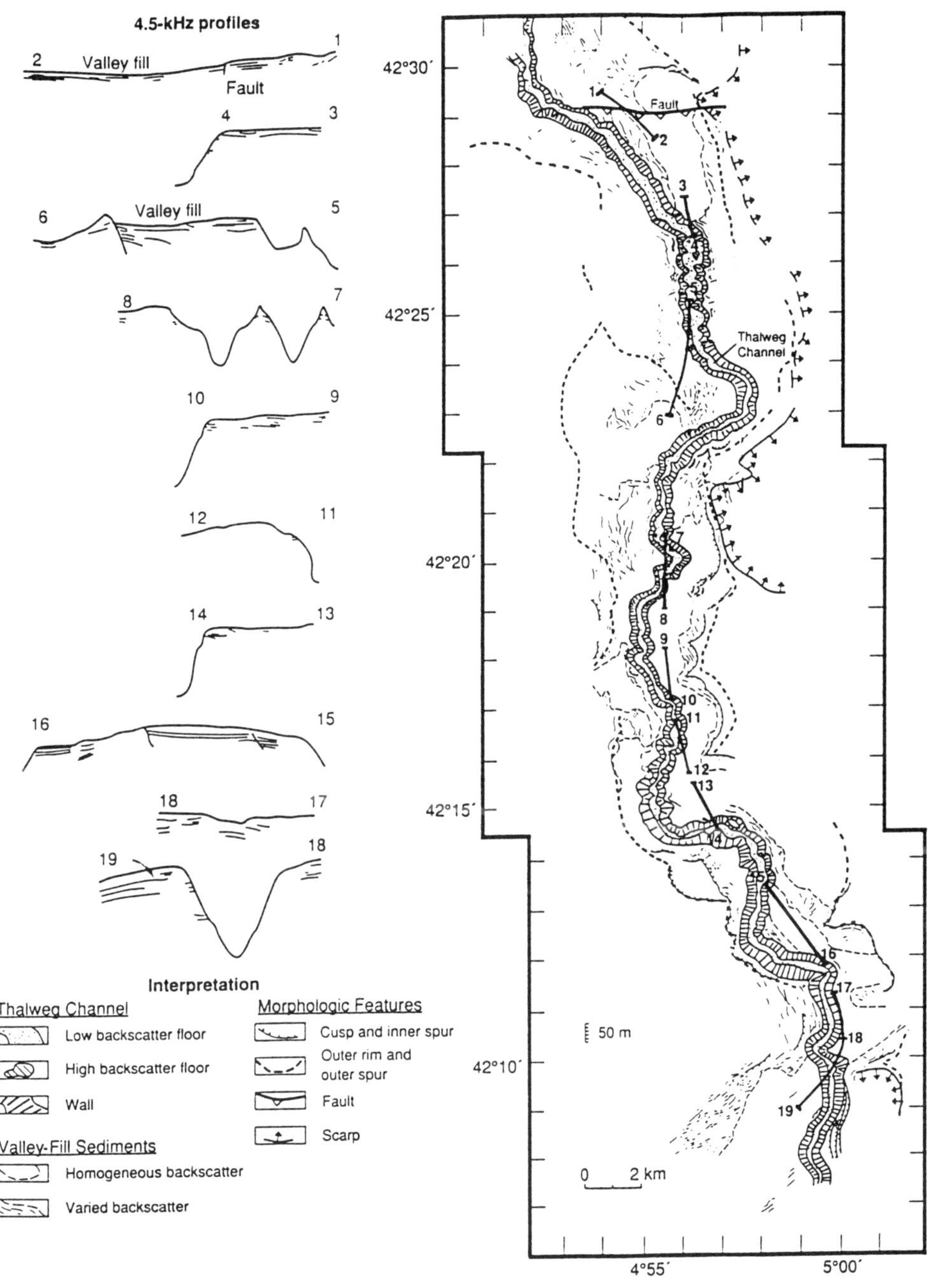

Figure 5.26. SeaMARC I interpretation of the upper Rhône fan channel and 4.5 kHz profiles (O'Connell *et al.* 1991).

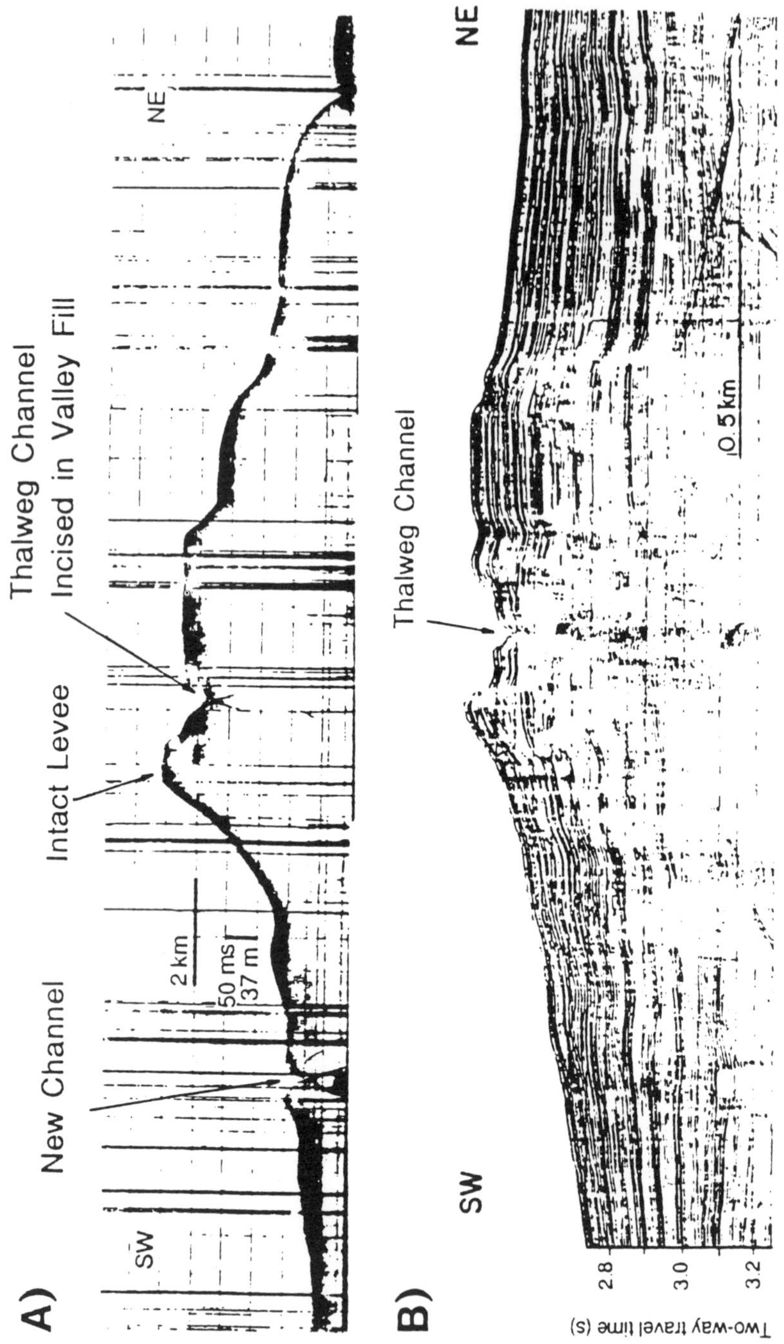

Figure 5.27. (a) 12 kHz profile of the Upper Rhône fan channel downstream of the avulsion point showing the abandoned thalweg and valley (and the aggradational character of this channel-levee complex), and the new channel. (b) Airgun profile upstream of the avulsion point showing the channel thalweg entrenched into the floor of the valley fill (O'Connell *et al.* 1991).

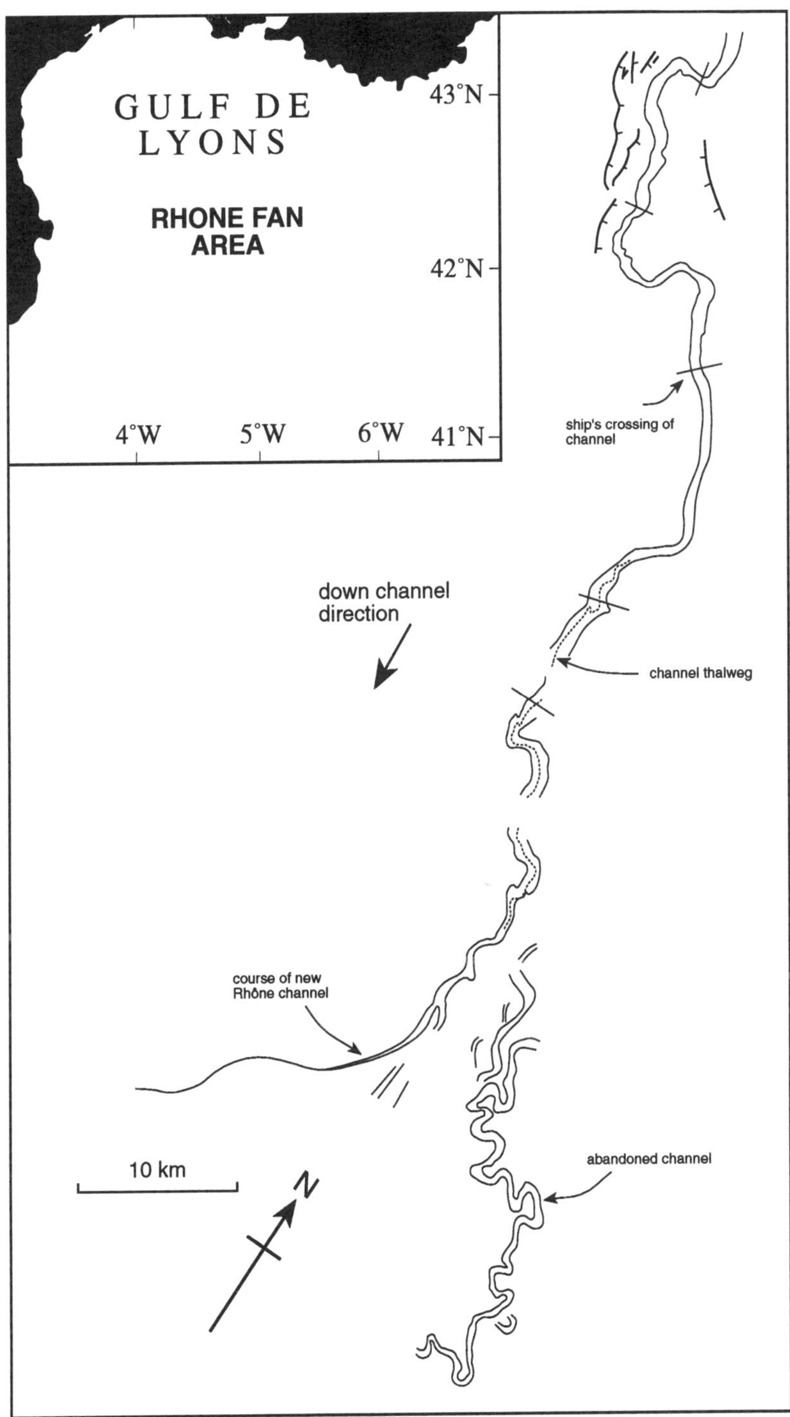

Figure 5.28. GLORIA interpretation of the Upper Rhône Channel showing the highly sinuous abandoned channel south-east of the most-recent avulsion point.

Chapter 6

Quantitative analysis of modern submarine channels

This chapter contains an expanded version of the paper published by Clark, Kenyon and Pickering (1992). In particular, graphical data for many modern submarine channel parameters are presented here for the first time, reproduced from the unpublished PhD thesis by Clark (1994).

Methodology

Many methods of quantitative analysis for channel planform geometry have been proposed by workers on fluvial systems. Flood and Damuth (1987) investigated the quantitative characteristics of the distributary channels on the Amazon Fan and concluded that certain planform similarities exist between the Amazon submarine channels and channels in large terrestrial river systems. The following section presents an analysis of submarine channel parameters from the submarine fans described above in order to examine: (i) the relationships between submarine channel parameters; (ii) the extent of the "fluvial-like" geometries of submarine fan channels; and (iii) the use of quantitative analysis in a more rigorous classification scheme for submarine channels and fans.

GLORIA data on the fan channels is taken from a variety of sources (see Table 6.1). The submarine fans were selected as representative of a range of fan depositional settings for which sufficient GLORIA data exist to allow essentially complete channel analysis.

The quantitative parameters used to analyse the planform geometry of submarine channels are similar to those employed by workers on fluvial systems. These parameters include sinuosity for particular sections of a channel or reach, down-channel changes in sinuosity, measurement of fan gradient, channel meander wavelength, radius of curvature, channel width, and channel depth (*cf.* Leopold & Wolman 1957, Schumm 1981).

Water depth (measured as from sea level to channel floor) and channel depth (mean levee height to channel floor) were gleaned largely from 3.5 kHz and precision echo sounder profiles, and in some instances supplemented with measurements taken from detailed bathymetric charts. The depth parameters were only obtainable when the ship tracks crossed a channel. The part of the channel between two successive crossings is defined as a "reach" (Flood & Damuth 1987). The parameters obtained from GLORIA data, are as follows: channel width (recorded at the start of each reach); channel meander wavelength, radius of

Channel	Location	Depth measurement method	Reference*
Amazon channel	Atlantic Ocean	3.5 kHz.	1
Arguello channel	West Coast USA	Bathym. maps	2
Astoria channel	West Coast USA	3.5 kHz.	3
Atlantis channel	East Coast USA	3.5 kHz.	4
Baltimore and Wilmington channels	East Coast USA	3.5 kHz.	4
Cascadia channel and tributaries	West Coast USA	3.5 kHz.	3
DeSoto channel	Gulf of Mexico	3.5 kHz., Bathym. maps	5
Grand Banks channels	East Coast USA	3.5 kHz.	4
Hudson channel	East Coast USA	3.5 kHz.	4
Indus channels	Indian Ocean	3.5 kHz.	6
Laurentian Fan channels	East Coast USA	3.5 kHz.	4
Mississippi channel	Gulf of Mexico	3.5 kHz., Bathym. maps	5
Monterey channel	West Coast USA	3.5 kHz.	2
Porcupine Seabight	Northeast Atlantic	3.5 kHz., Bathym. maps	7
Rhône channel	Mediterranean	3.5 kHz., Bathym. maps	8
Umnak and Pochnoi channels	Bering Sea Basin	Bathym. maps	7

*1—Damuth et al. (1983); 2—EEZ-Scan 84 Scientific Staff (1986); 3—Hampton et al (1989); 4—EEZ-Scan 87 Scientific Staff (1990); 5—EEZ-Scan 85 Scientific Staff (1987); 6—Unpublished data; 7—Bering Sea EEZ-Scan Scientific Staff (1991); 8—O'Connell et al. (1991).

Table 6.1. Sources of data for the quantitative study of submarine channels (Clark *et al.* 1992).

curvature (the average values for any meander loops in each reach), valley length (the straight line distance between successive crossings), and channel length (the distance measured down channel between successive crossings).

Using these parameters, valley slope, channel slope, sinuosity (ratio of channel length to valley length), and width/depth ratio for individual reaches can be calculated.

The errors in calculating the depths from the 3.5 kHz profiles have been estimated to be of the order of up to 10 m. This error is largely due to the channel floor reflection being obscured by the side echoes from the channel walls. Where 3.5 kHz and precision echo sounder data are unavailable, the depths have been measured from bathymetric charts, and these measurements may have errors up to a maximum of 40 to 50 m. for the depth to seafloor and about 20 m. for channel depths. In addition, the measurements of channel parameters may not reflect the true "flow-forming" channel geometries, because many of them have become less active during the Holocene high sea-level stand and there may be modifications such as channel-wall sliding.

It is important to note that the parameters derived from the GLORIA data are applicable to each particular reach, where as the parameters derived from the 3.5 kHz data are obtained from the start and finish of each reach. "Mid-reach" parameters are plotted against the "start-of-reach" values in the following analysis (i.e. channel depth at the top of each reach is assumed to remain constant for the length of that reach).

The unavoidable arbitrary quantifier required for this method of analysis is the length of each reach. This parameter affects the calculations of sinuosity and meander radius of curvature and wavelength since some of the smaller reaches contain only partial meander loops, if any.

Table 6.2 summarises some of the important quantitative characteristics of the submarine channels used in this study. Since submarine channel width and depth vary along the length of a channel, the width, depth and aspect ratios for the greatest cross-sectional area are used to characterise each channel.

channel system	Arguello	Astoria	Atlantis	Baltimore-Wilmington	Bering Channel	Cascadia Channel	DeSoto Channel	Grand Banks Valley	Hudson
location	West Coast USA	West Coast USA	East Coast USA	East Coast USA	Bering Sea	West Coast USA	Gulf of Mexico	East Coast USA	East Coast USA
scale (at greatest area of channel cross-section) — width (km)	1.5	2.9	2.55	6.5	max. width 3.6	5.6	0.6	9.4	3.4
depth (m)	302	165	300	195	-	165	17	465	500
approximate aspect ratio	5	18	8.5	33	-	34	35	20	7
maximum sinuosity	1.54	1.07	1.14	1.07	1.22	1.41	2.20	1.07	1.48
sediment supply mechanism	Canyon	Columbia River and Astoria Canyon	Canyon	Canyon	Volcanic island arc	Canyon	?	Continental slope	Canyon
reference	EEZ-Scan 84 Scientific Staff 1986, 1988	Kulm et al. 1973 Nelson & Nilsen 1974 Nelson 1985 Hampton et al. 1989	EEZ-Scan 87 Scientific Staff 1990	EEZ-Scan 87 Scientific Staff 1990 Twichell et al. 1980	Bering Sea EEZ-Scan Scientific Staff 1991 Kenyon & Millington 1995	EEZ-Scan 84 Scientific Staff 1986, 1988 Hampton et al. 1989	EEZ-Scan 85 Scientific Staff 1987 Twichell et al. 1991	EEZ-Scan 87 Scientific Staff 1990 Hughes Clarke et al. 1990 Masson et al. 1985	EEZ-Scan 87 Scientific Staff 1990 Twichell et al. 1980

channel system	Indus "A" Channel	Indus "a" channel	Laurentian	Mississippi	Monterey	Pochnoi	Porcupine	Rhone	Umnak Channel
location	Indian Ocean	Indian Ocean	East Coast USA	Gulf of Mexico	West Coast USA	Bering Sea	Northeast Atlantic	Northwest Mediterranean	Bering Sea
scale (at greatest area of channel cross-section) — width (km)	11	0.75	13	7.5	2	Max. width 3.9	2.6	1.0	Max. width 6.2
depth (m)	410	96	330	22	884	-	370	120	-
approximate aspect ratio	27	8	40	340	2.3	-	7	8	-
maximum sinuosity	3.13	2.45	1.04	1.74	1.60	1.07	1.15	1.48	1.13
sediment supply mechanism	Indus River and Canyon	Indus River and Canyon	Continental slope and St. Lawrence River	Mississippi River	Monterey Canyon	Volcanic island arc	Celtic Shelf	Rhone River and Canyon	Volcanic island arc
reference	Kolla & Coumes 1985, 1987 Kenyon et al. 1995	Kolla & Coumes 1985, 1987 Kenyon et al. 1995	Masson et al. 1985 EEZ-Scan 87 Scientific Staff 1990 Hughes Clarke et al. 1990	Garrison et al. 1982 Pickering et al. 1986b EEZ-Scan 85 Scientific Staff 1987 Twichell et al. 1991	Menard 1955 EEZ-Scan 84 Scientific Staff 1986, 1988 Gardner et al. 1991 Masson et al. 1995	Bering Sea EEZ-Scan Scientific Staff 1991 Kenyon & Millington 1995		Bellaiche et al. 1893, 1986a O'Connell et al 1991	Bering Sea EEZ-Scan Scientific Staff 1991 Kenyon & Millington 1995

Table 6.2. Summary table of the modern submarine channels included in this study.

Channel dimensions

Channel depth vs. down-channel distance

The data from some of the channels is not quite as comprehensive as that for the Amazon Fan channels (Flood & Damuth 1987), but similar trends can be seen (e.g. the Arguello and the Monterey channels where Canyon data have been included in the results) (Figure 6.1).

The observed trend (an initial increase in depth followed by a gradual decrease in depth) is similar to that observed by Menard (1964), and Komar (1973). Komar modelled data for deep sea channels using Chezy- type equations to obtain curves similar to those observed in deep sea channels. He concluded that this observed trend in channel relief was due to 'continuity of turbidity current flow', causing a thickening of the turbidity current as it flows down progressively shallower gradients, and hence a consequential deepening of the channel. The 'continuity of flow' model may have a significant control on channel dimensions. The increasing channel depth trends (shown on Figure 6.1) correspond to portions in which the main channel is collecting sediment from numerous tributaries and receiving sediment from the distal Astoria Fan, rather than distributing it. The Cascadia Channel also underwent major incision due to its tectonic control by the Blanco Fracture Zone. (Carter 1988, Komar 1973).

These results are consistent with the fact that as turbidity currents decelerate along their flow path, i.e. down-fan, there is a tendency for the flow thickness to initially increase, prior to their eventual cessation. Exceptions to this pattern may occur when there is a tectonic lowering of the base level at the depositional end of a channel to cause renewed channel incision. Additionally, the role of tributaries in supplying sediment into a particular conduit will have effects on the down-channel changes in morphology in a complex manner.

Channel width vs. down-channel distance

The graph in Figure 6.2 shows similar trends to those of the channel depth data, largely due to the reasons mentioned above. Note that the width data for the Amazon Fan channels does not extend back as far as their depth data and so the 'peaked' trend cannot be confirmed, but again, the Arguello and the Monterey as well as the Aleutian Basin channel widths show an initial width increase followed by a gradual decrease. The upturn in the curves at the ends of the Mississippi and Aleutian Basin channels are a result of bifurcation of the channels resulting in braid type lobe deposition. Strictly, the dimensions of the braided channels should have been measured rather than the width of the 'braid valley', which would have resulted in a continued narrowing trend.

Cross sectional area vs. down-channel distance

From the results of the width and depth data it appears that channel cross sectional area would be a better measure of systematic down-channel changes in channel dimensions.

Channel cross-sectional area has been approximated to the product of width and depth. Note that cross sectional area is plotted on a logarithmic scale (Figure 6.3). Unfortunately, the combination of the two incomplete datasets (width and depth), has resulted in a loss of potential data. Because of this only the DeSoto, Hudson, Grand Banks, and Laurentian channels show the 'peaked' trend in cross sectional area, but it is assumed from the conclusions drawn above that the other channels would show similar trends.

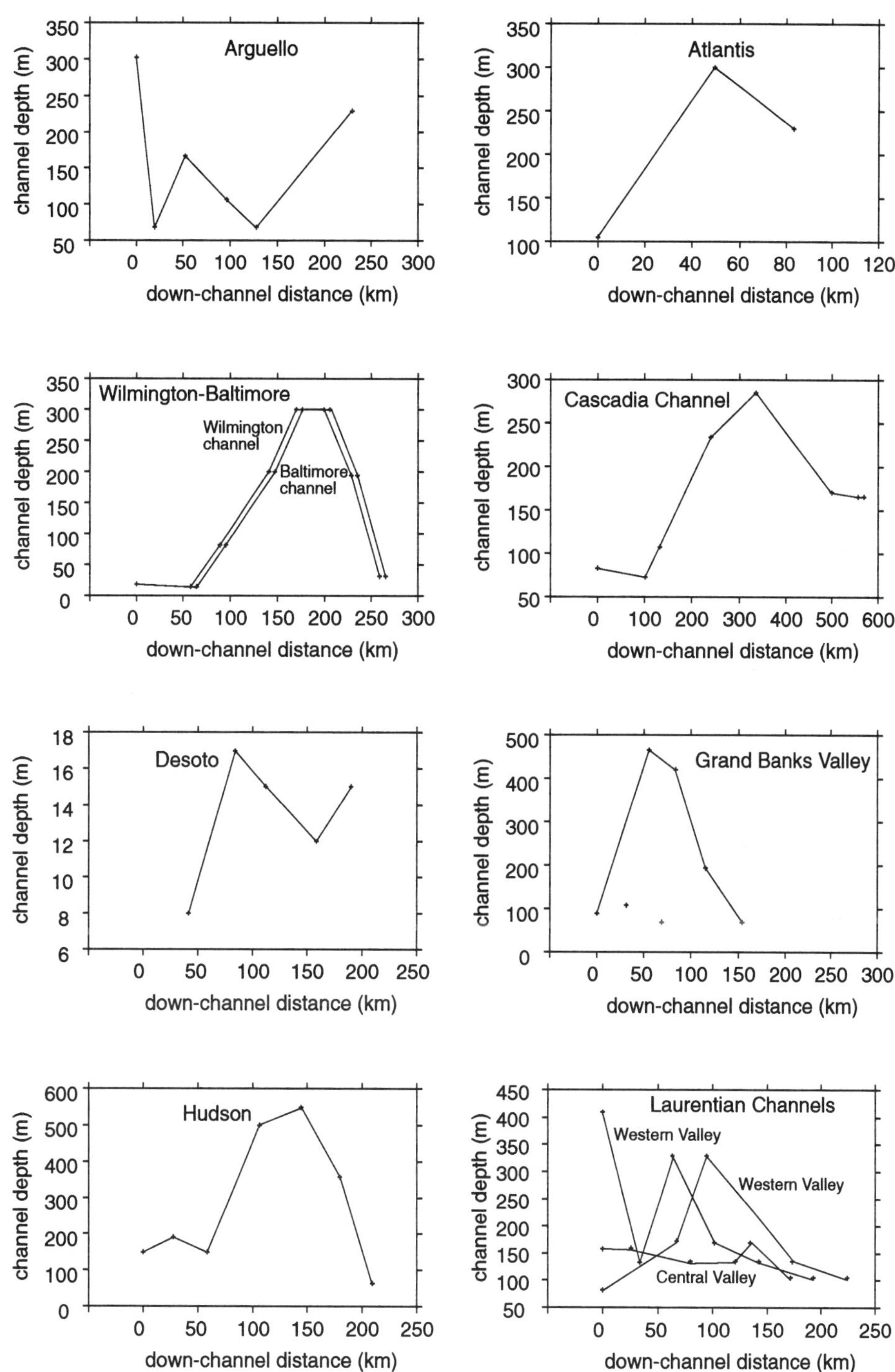

Figure 6.1. Plot of channel depth against down-channel distance

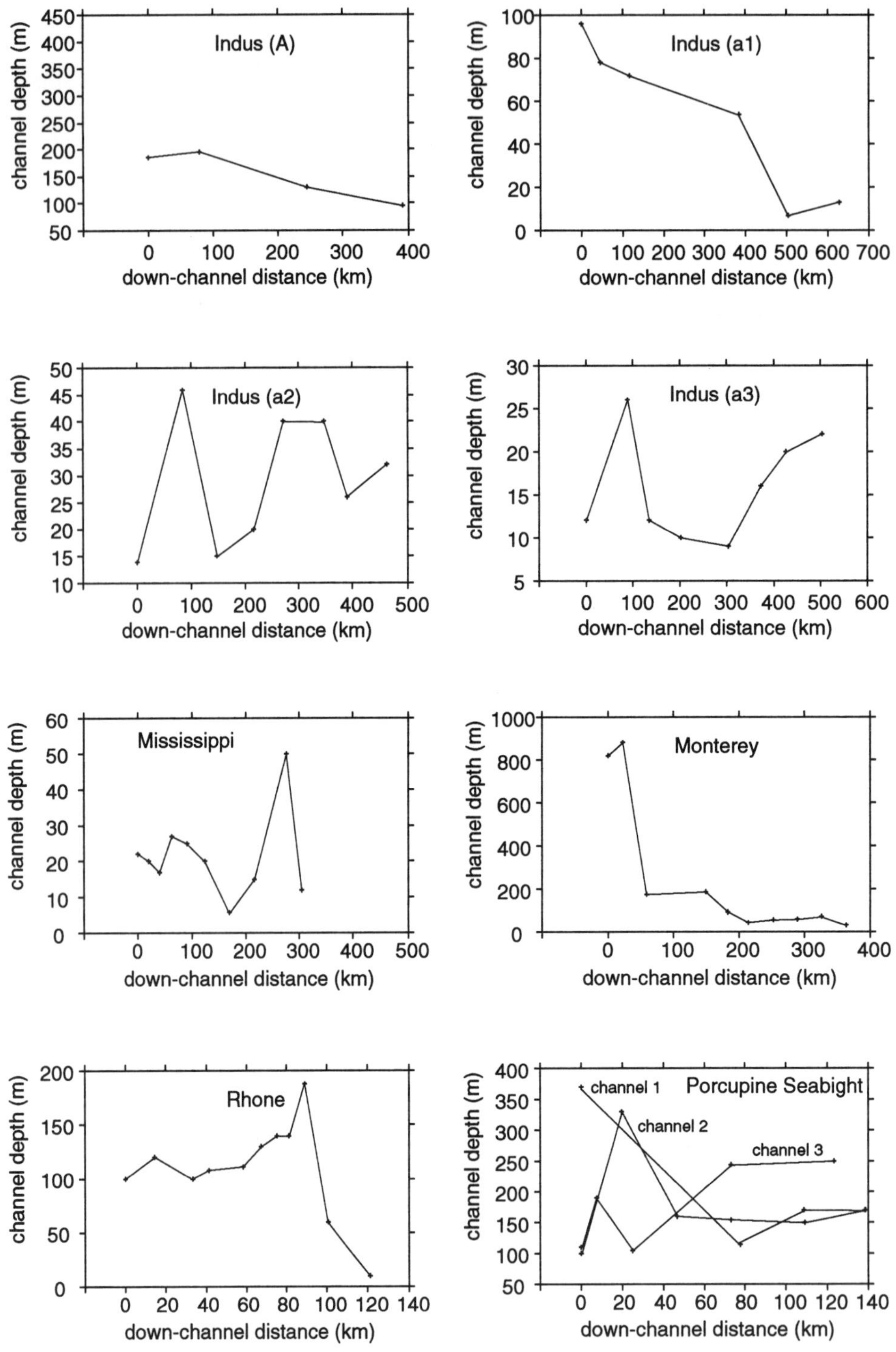

Figure 6.1. Continued

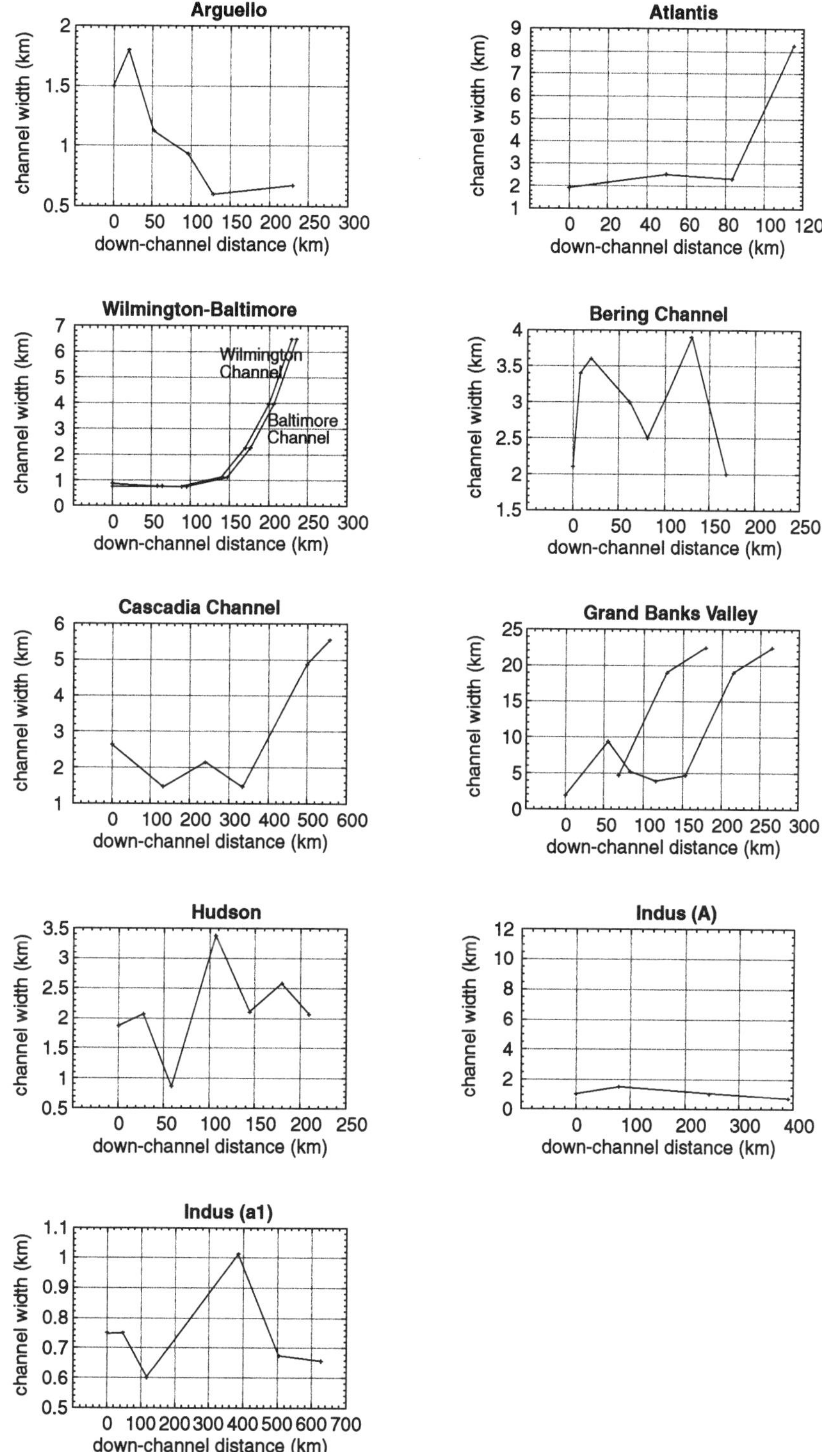

Figure 6.2. Plot of channel width against down-channel distance.

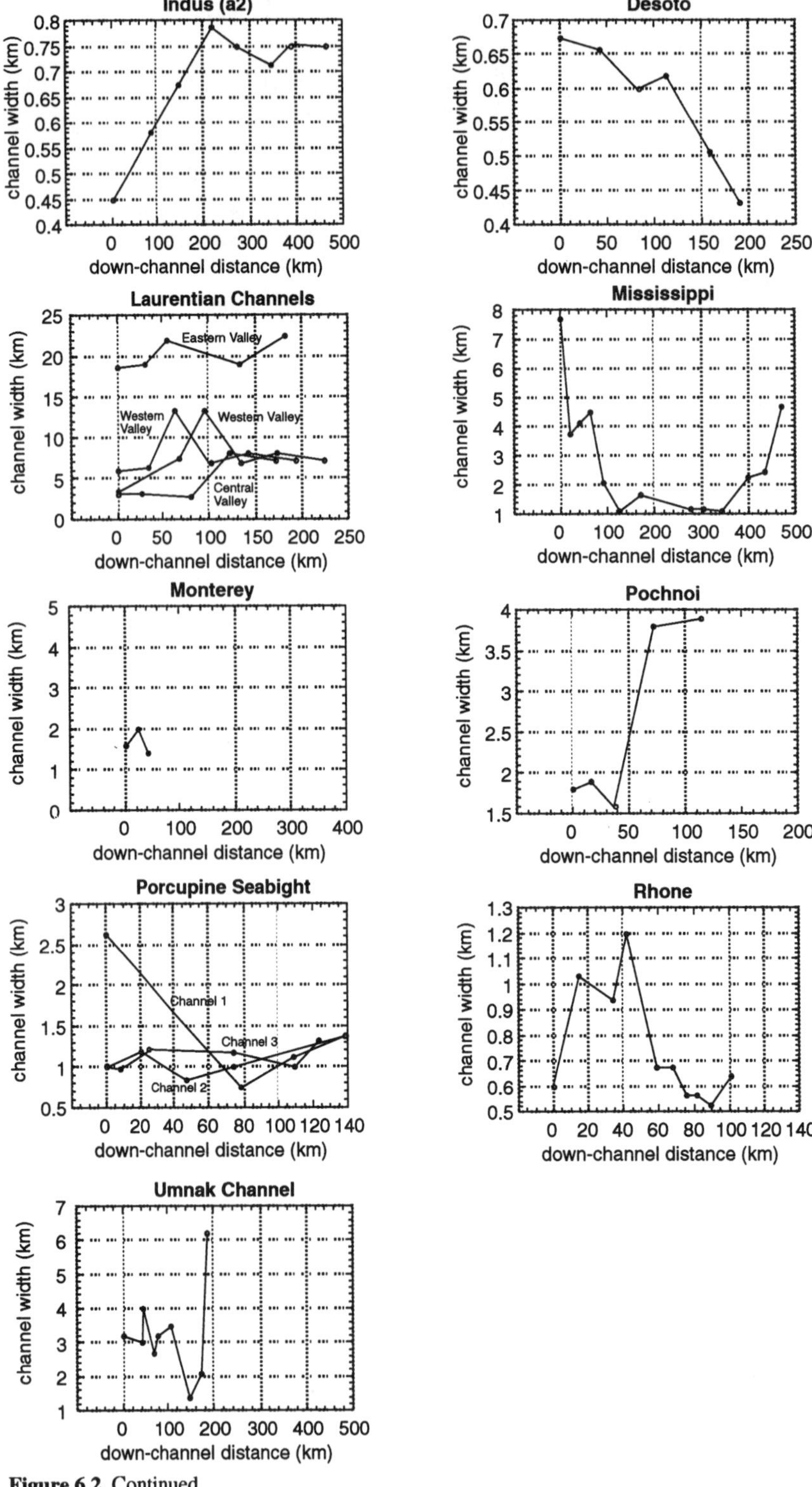

Figure 6.2. Continued

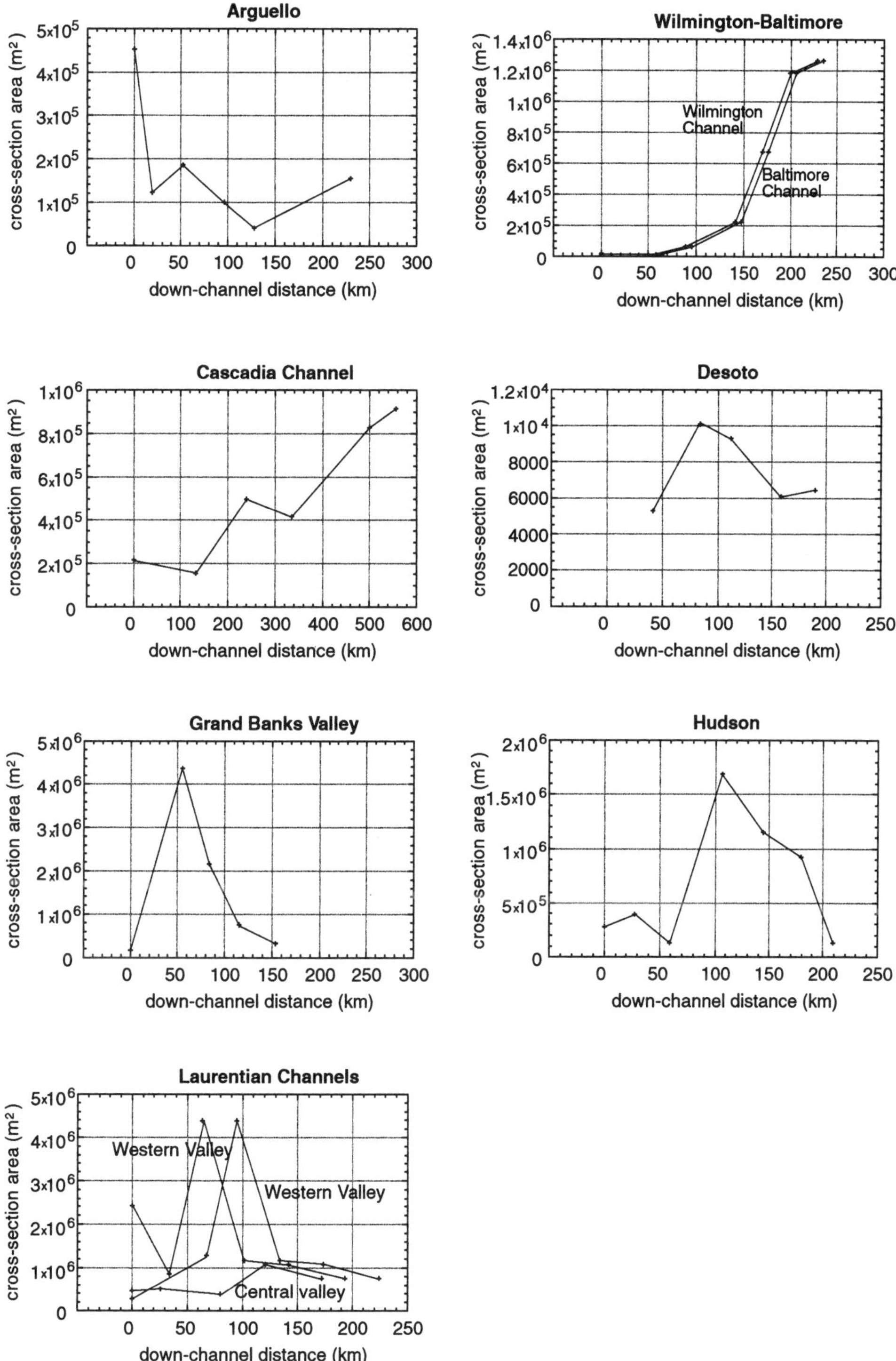

Figure 6.3. Graphs of channel cross-section area against down-channel distance.

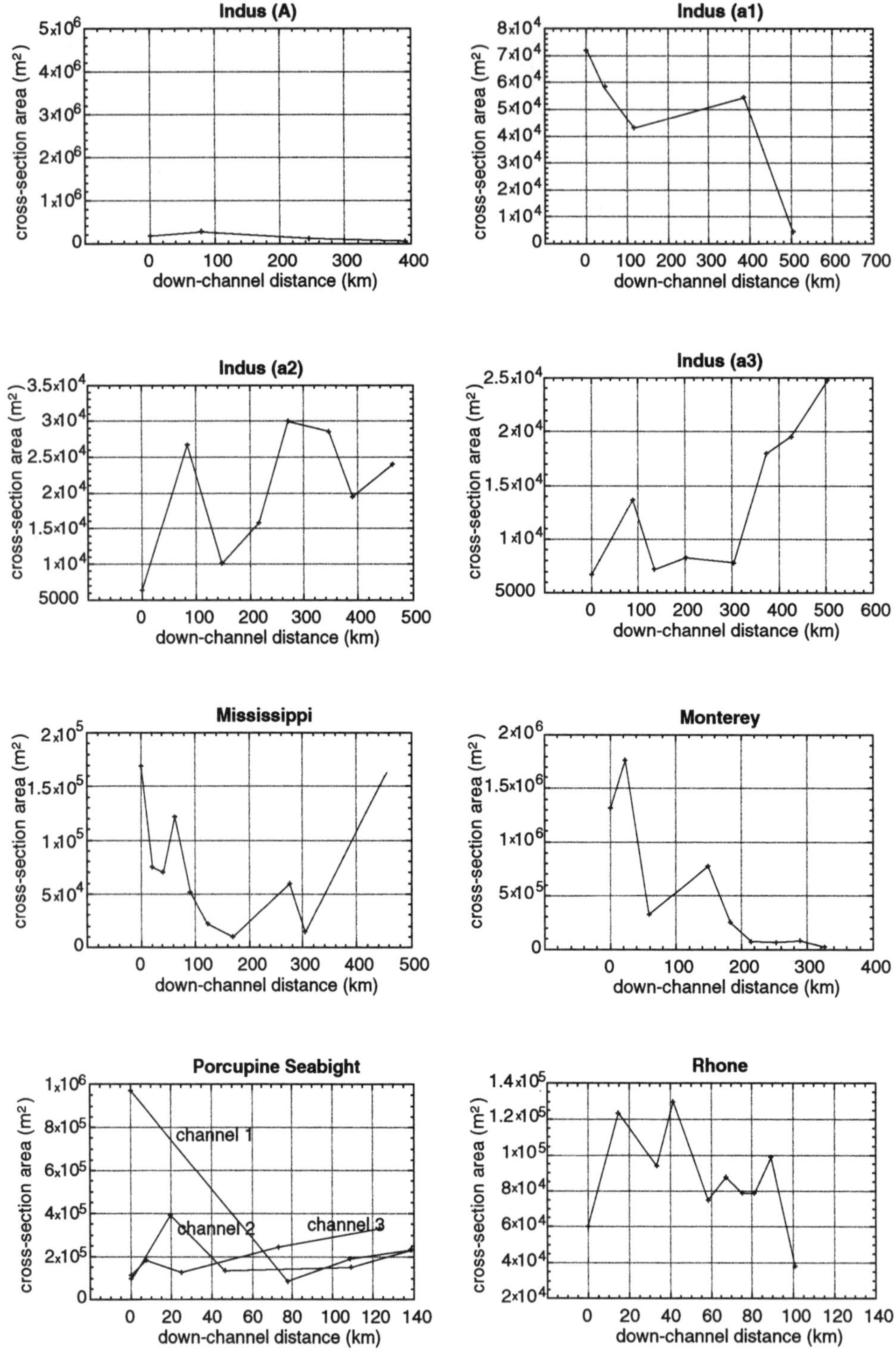

Figure 6.3. Continued.

Channel width vs. channel depth

The results of width and depth measurements can be plotted in this form to reveal the nature of channel geometry (Figure 6.4). This graph is important in the study of ancient channel examples, since width and depth can be obtained from measurements in the field and from borehole and seismic data. The graph shows a scatter of data points reflecting a large variance in modern deep-marine channel aspect ratios. It is interesting to note that the selected examples of ancient deep-marine channel width and depth measurements, plot in a field of smaller dimensions than that of the modern submarine channels. This may be due to underestimating ancient channel dimensions by only recognising sandy fill parts of channels, or it may be due to some elementary disparity between modern channels and the type of ancient channels that get preserved.

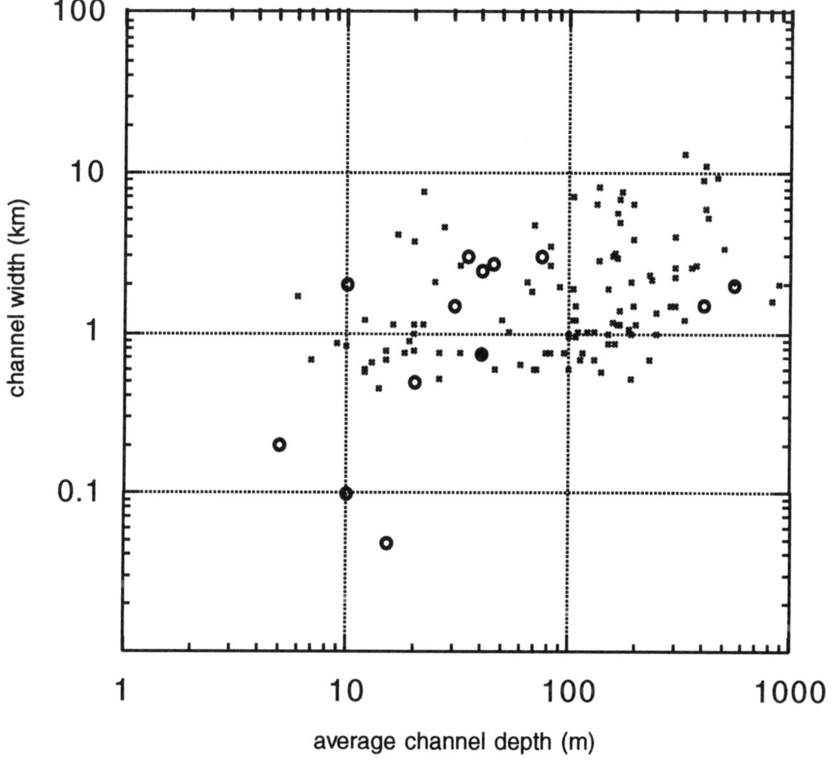

Figure 6.4. Plot of submarine channel width against depth for various modern channels. Data for ancient submarine channel and canyon fill are also plotted for comparison.

Meander geometry

For the purpose of this study, the wavelength of meander loops and their radii have been averaged to give a mean meander wavelength parameter for each reach. Meander wavelength parameters can then be plotted against channel width and radius of curvature on log/log plots. Leopold and Wolman (1960) collected data from many different river systems in the U.S. and found that a good correlation exists between these two parameters. Likewise, Flood and Damuth (1987) demonstrated that very similar relations exist on the Amazon Fan. Bellaiche *et al.* (1986b) analysed the meanders of part of one of the Rhône Fan channels and found the morphology to be more similar to that of the Martian channel Nirgal Vallis than to that of fluvial channels. The "average" relations between such meander parameters for all the submarine channels studied in our analysis lie closer to those obtained from fluvial data (see Figures 6.5, 6.6)

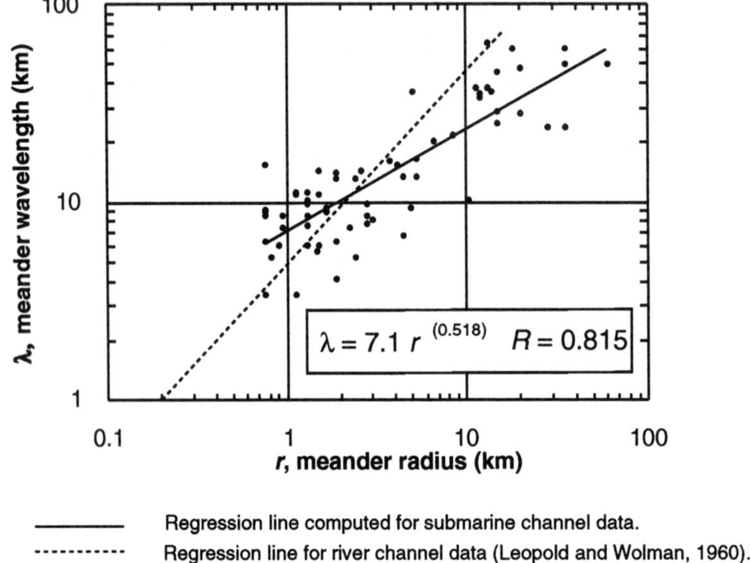

Figure 6.5. Plot of meander wavelength against radius of meander curvature for all submarine channel data.

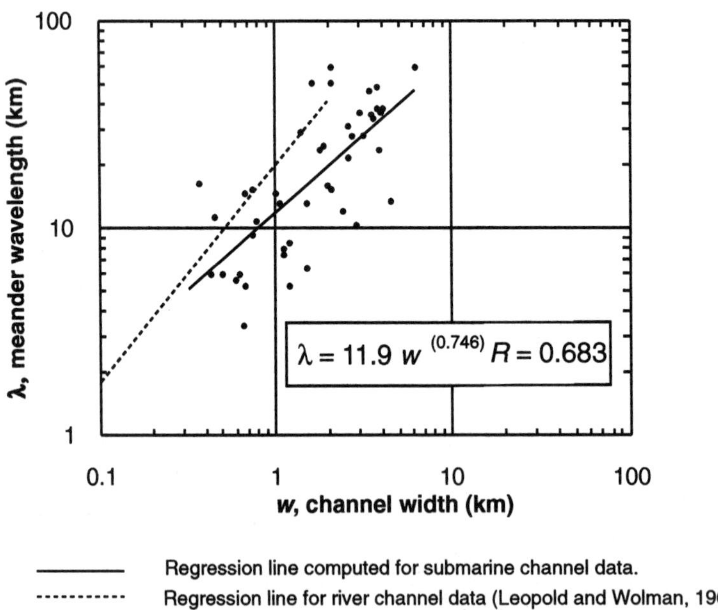

Figure 6.6. Plot of meander wavelength against channel width for all submarine channel data.

Meander wavelength vs. meander-loop radius

The data for meander wavelength versus meander-loop radius is plotted on a graph with logarithmic scales (Figure 6.5). The data shows a relatively consistent relationship between wavelength and radius of meanders. The empirically derived relationship for their river data is shown in Figure 6.5. The deep-sea channel data does not exactly fit the river channel data but there is a relatively good regression line with a correlation coefficient of 0.815.

Meander wavelength vs. channel width

The graph of meander wavelength vs. channel width gives a regression line with a correlation coefficient of 0.683, but again this is different to the correlation for the river channel data (see Figure 6.6). The differences shown between the river channel data and the deep sea channel data may be due to the inaccuracy of measuring deep-marine channel meander wavelength and radius. Both these parameters were obtained from a calculation of the average value for all the meander loops for a given reach, where as the Leopold and Wolman (1960) data was derived from measurements of each individual meander loop.

Sinuosity variation

The variation of channel sinuosity with increasing gradient is an important relationship that has been observed in rivers and in flume tank experiments (Schumm & Khan 1972, Schumm *et al.* 1972). Langbein and Leopold (1966) showed that the shape of a meander represents an adjustment of depth, velocity and slope to minimise the variance of shear and frictional resistance. Figure 6.7 shows the peaked sinuosity trend found by Schumm and Khan (1972), and Schumm *et al.* (1972), for flume-tank channels and also for the Mississippi River. The explanation given by Schumm and co-workers for this trend is that as slope increases, channel sinuosity will increase in such a way as to maintain an optimum "channel slope" suitable to accommodate the volume of flow and sediment load in the channel. This process will

continue until a "threshold slope" is reached, after which the channel will be searching for a more direct course downslope, resulting in a rapid decrease in sinuosity. Studies of submarine channel sinuosity and channel slope generally show good correlations, with trends that can be shown to fit with the scaled curves obtained from river and flume tank data. Generally, the scatter of data around the curves obtained from submarine channels is similar to that obtained from fluvial channels (Figure 6.8). The good data fit may be attributed to the lack of heterogeneity in the 'bed rock' into which submarine channels are cut compared to the heterogeneity in the bed rock over which terrestrial rivers flow, despite the errors and difficulties incurred in measuring submarine channels.

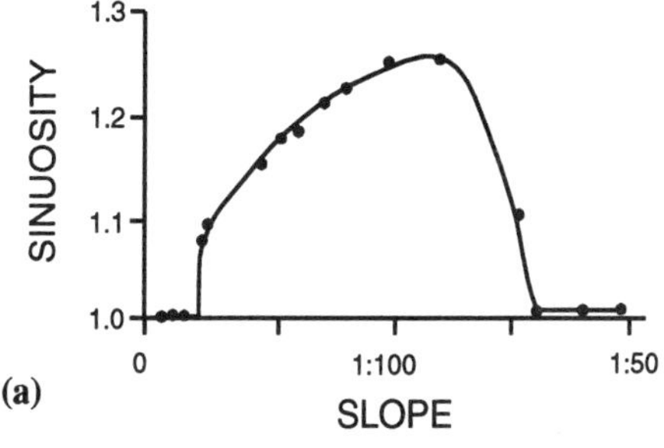

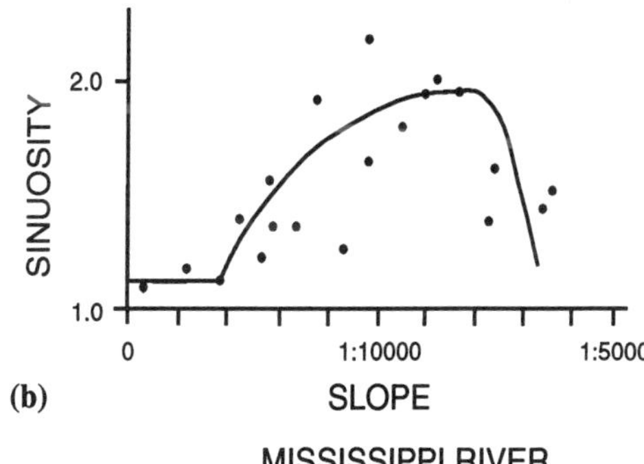

Figure 6.7. (a) Graph of channel sinuosity against slope for flume tank channels (Schumm & Khan 1972). **(b)** Graph of channel sinuosity against slope for the Mississippi River (Schumm et al. 1972). Both graphs show a peaked trend in maximum sinuosity.

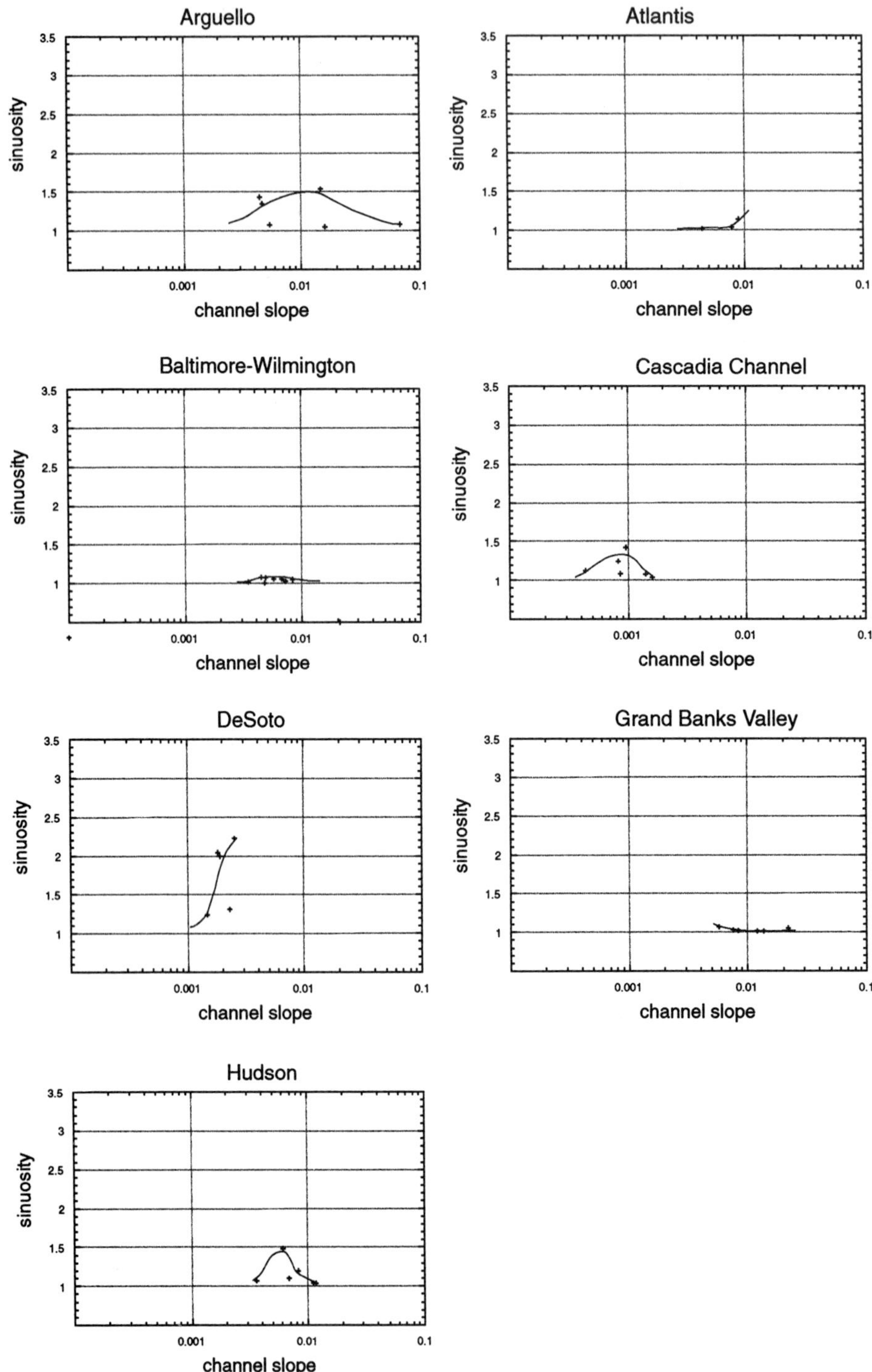

Figure 6.8. Graphs of deep-marine channel sinuosity against channel slope, with lines of best fit showing slope and sinuosity relationships.

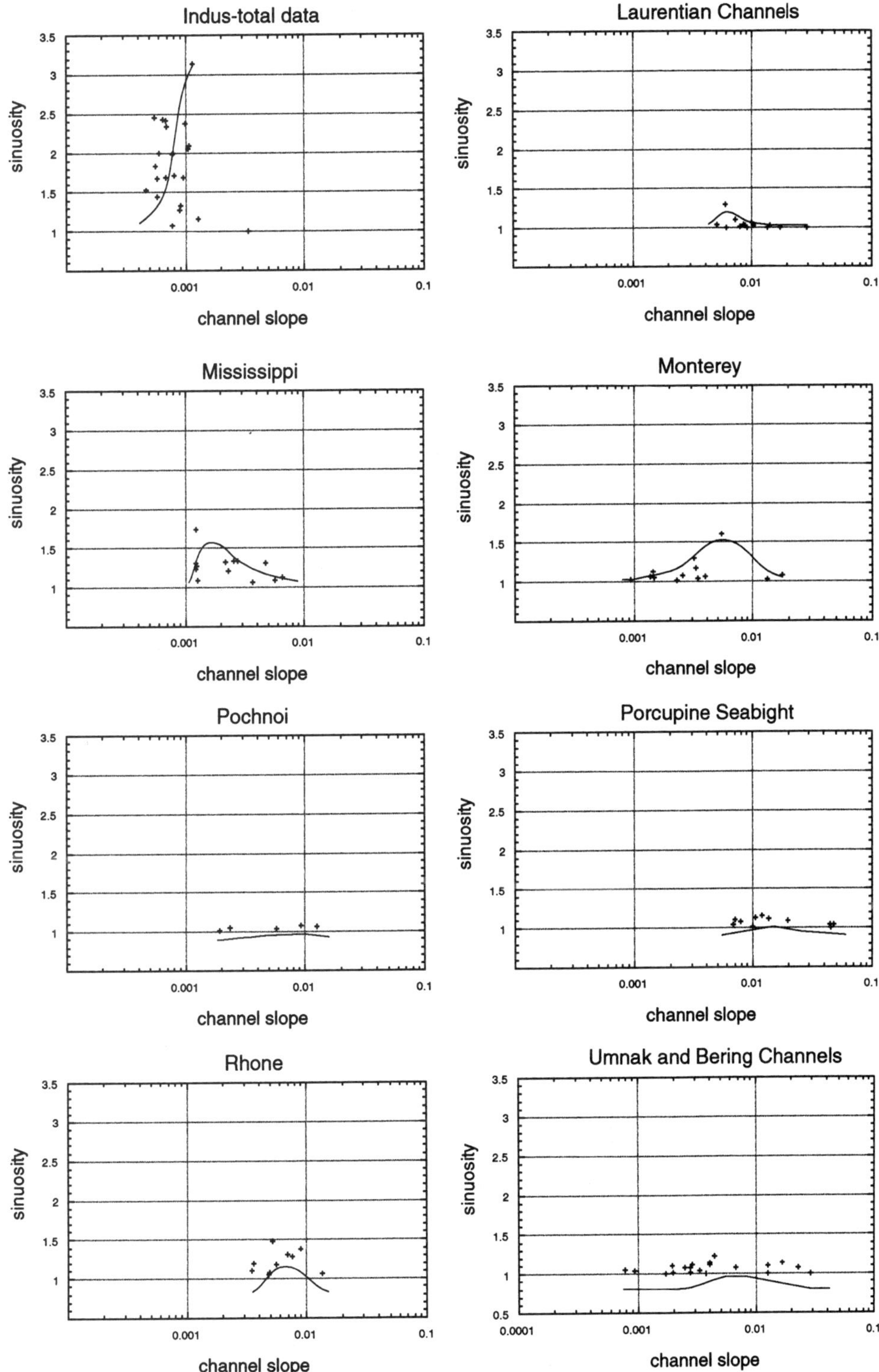

Figure 6.8. Continued.

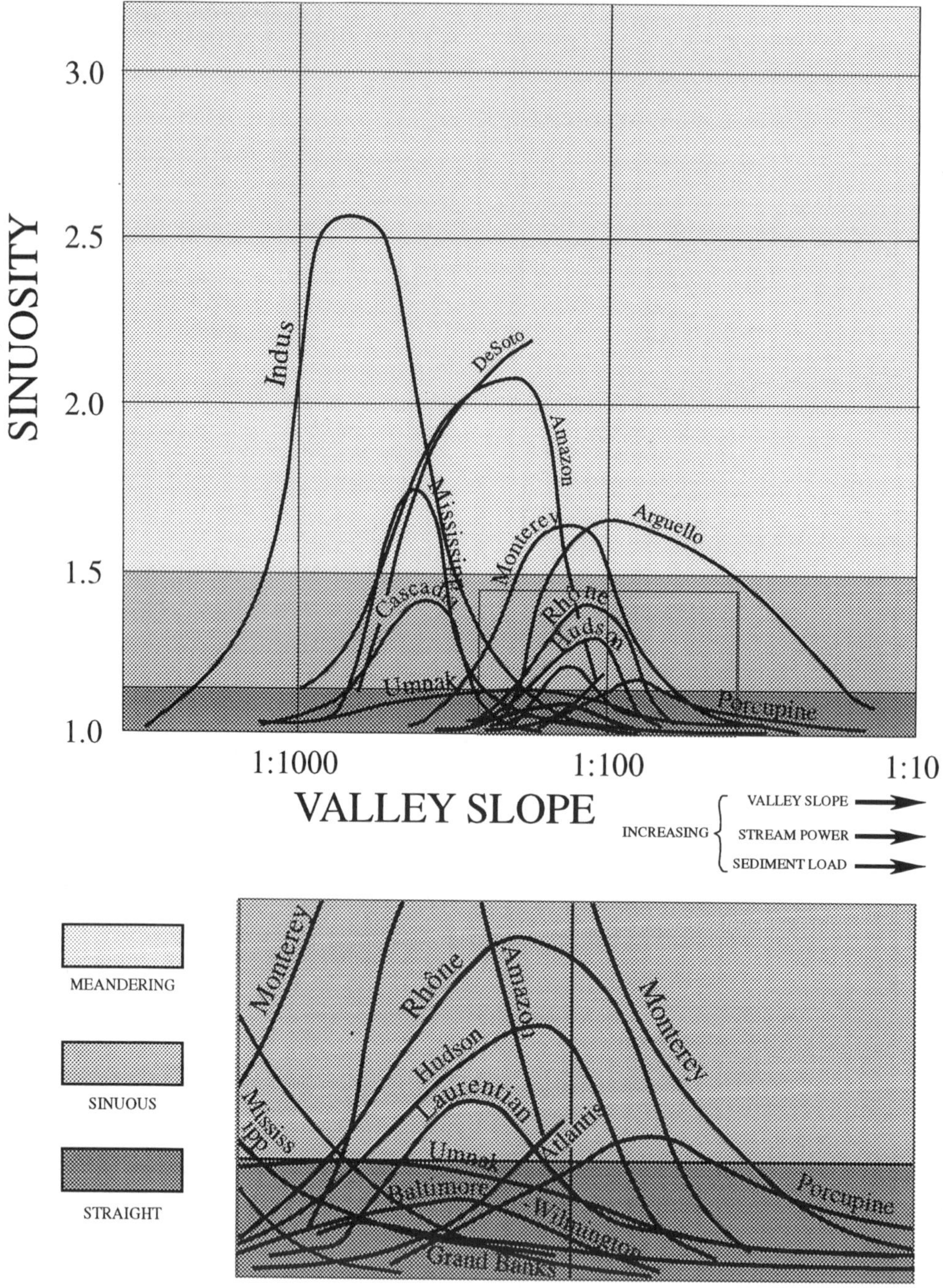

Figure 6.9. Sinuosity and valley slope relationships for selected submarine fan channels showing trends similar to those of flume tank and fluvial data shown in Figure 6.8. Amazon Channel data taken from Flood and Damuth (1987). Lower box shows the detail from the main graph. After Clark, Kenyon & Pickering (1992).

Figure 6.9 shows the best-fit curves obtained from sinuosity and valley slope relations for all the fan systems in this study. A peaked sinuosity trend can be shown for most of the submarine channel data, showing different "threshold" sinuosity values and their corresponding valley-slope values in different fan systems. There is a general trend from the most sinuous systems (e.g. the Indus Fan channels) obtaining a "peak" sinuosity on relatively low gradients (about 1:400) to the least sinuous systems (e.g. the Porcupine Seabight channel system) which reach their "peak" sinuosity on steeper slopes (about 1:80).

Schumm (1981) investigated the response of channel planform to sediment load, and showed that water discharge, flow character, volume, and calibre of sediment determine fluvial channel morphology to a major extent. These factors are also likely to affect submarine channel morphology. Figure 6.10 shows the effect of valley slope or sediment load for different sediment load types in streams. The data suggest that for a coarser bed-load, channels tend toward a lower sinuosity "peak" value than those dominated by suspended-load sediment.

Intuitively the type of sediment transported in channels should be an important contributory factor in determining the morphology of channels, as well as valley slope. The exceptions to this trend can perhaps be attributed to factors such as flow type (e.g. occurrences of major debris flows), seafloor topography, and tectonic events affecting the channel course. For example, the Arguello Channel is sited on thinly covered oceanic basement (EEZ-Scan 84 Scientific Staff 1988),

and much of the "sinuosity" measured in this analysis is likely to be attributed to control from the rather irregular pre-existing seafloor topography.

The Cascadia Channel is predominantly tectonically controlled and, in its proximal part, crosses the lower slope of an accretionary prism, resulting in localised confinement to discrete terrace levels (Karl *et al.* 1989), and in its distal segment, the channel route and plan-form geometry is governed by uplift and subsidence in the Blanco Fracture Zone (Embley 1985).

The Mississippi channel data shows an anomalously low peak sinuosity obtained on relatively low gradients. The reason for this is as yet unclear but Twichell *et al.* (1991) have suggested that large volumes of debris flows have been transported the length of the Mississippi Channel. Channel sinuosity may have been affected by such large volumes of debris flows.

Summary of quantitative analysis, and implications for classification of fans

As has been observed by other researchers, some submarine channels exhibit sinuous planform geometries that are comparable to those of fluvial channels. This is probably a result of similar physical processes operating in submarine and fluvial systems. Results from an analysis of meander wavelength, radius, and channel width relation give similar results to those derived for the Amazon Fan by Flood and Damuth (1987). The results of this study show that, in general, the planform geometry

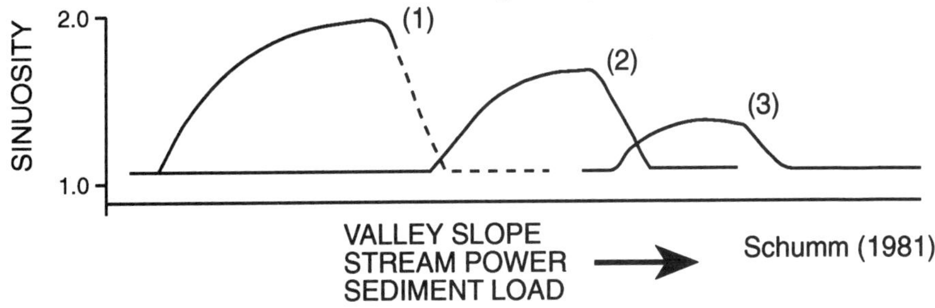

Figure 6.10. The effect of water discharge, flow character, slope gradient, volume and calibre of sediment on fluvial channel sinuosity. Curves show (1) suspended sediment load, (2) mixed load, (3) bed-load stream. (Schumm 1981).

of submarine channels is largely comparable in size and shape to some of the world's largest fluvial systems, and demonstrate similar relations between their diagnostic parameters. There is, however, less of a correlation for the trends observed in submarine channels compared to various subaerial river channels, largely due to the errors in extracting measurements from the data.

Current classifications of submarine fans concentrate on relating fan shape and sediment calibre (e.g. Pickering 1982b, Stow 1985, Reading 1991, Reading & Richards 1994). For relatively small fans, especially those most likely to be preserved in the rock record, fan shape is generally a secondary consideration because it is determined largely by basin shape which is, in turn, controlled by the tectonic environment. The use of more objective criteria is required for the classification of modern fans.

Channels are a common morphological element of fan systems, and their morphology reflects the nature of sedimentary processes active on the fan. By measuring channel parameters, a more quantitative and objective classification of fan systems can be formulated, from which depositional models may be constructed.

A quantitative approach to submarine-fan channel geometry allows a potentially more rigorous method of classifying fans. This study has identified the important factors that control the development of submarine channel systems: fan gradient (valley slope), sediment load type and calibre, and possibly flow type (e.g. the Mississippi Channel), and the density of turbidity currents. In fan systems, where factors such as basin shape are likely to exert a strong control on fan and channel morphology and location, as in the Cascadia Channel and Arguello Channel, then the relations and trends discussed may not hold true, or at least they may require modification.

As fan gradient decreases, sinuosity increases to a maximum "peak" sinuosity and then subsequently decreases. The peak value is a unique quantifiable parameter for each fan system. Using such parameters, a simple fan channel classification scheme can be constructed and applied to a fan classification

whereby fans are basically classified by the nature of the physical processes that control the morphology of their channels. Fans can be classified between two end members: high-sinuosity, low-gradient systems (e.g. the Indus Fan) and low-sinuosity, high-gradient systems (e.g. the Porcupine Seabight system). Deviations from this simple classification can be attributed to major tectonic controls on fan channel morphology or significant differences in flow type (Figure 6.11).

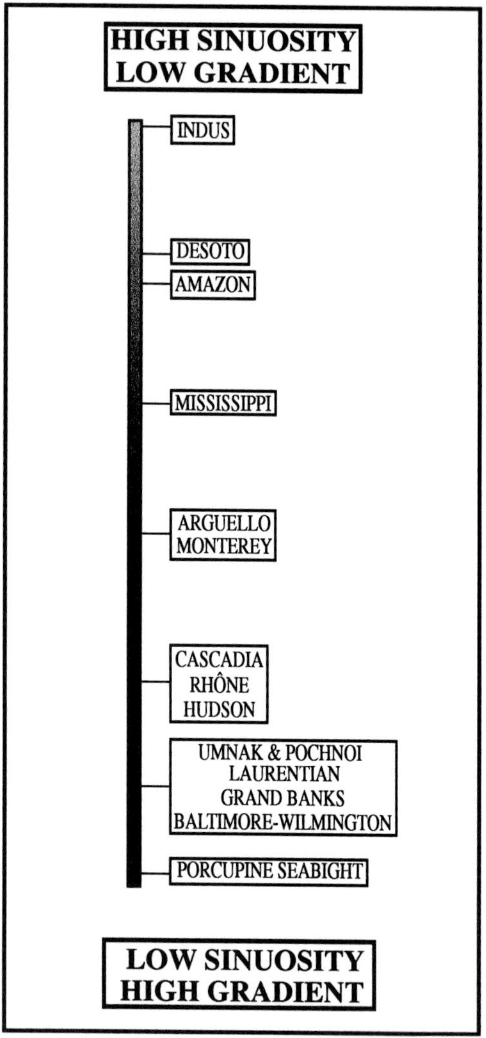

Figure 6.11. A classification scheme for submarine channels and/or submarine fans based on their channel planform geometry.

93

Part III

Ancient Systems

Chapter 7

Architecture of ancient channel-levee complexes

Eocene Hecho Supergroup, submarine channels, south-central Pyrenees, Spain

The deep-marine deposits of the Eocene Tremp-Pamplona Basin are collectively known as the Hecho Supergroup (Mutti *et al.* 1972). The Hecho Supergroup may be considered as a turbidite complex sensu Mutti and Normark (1987), where they define it as "a turbidite complex refers to a basin-fill succession and is composed of several turbidite systems that are stacked one upon the other". Hecho Supergroup sedimentation occurred in the central and western sectors of the basin during the Castisent, Santa Liestra and Campodarbe Group times (Figure 7.1). The deposits of the Montana Group are the stratigraphic equivalents of the deep-marine Hecho Supergroup, and consist of fluvial and deltaic successions in the eastern sector of the Tremp-Pamplona Basin (Mutti *et al.* 1972).

The Hecho Supergroup has been divided into several turbidite systems recording different stages of basin fill of a long-lived turbidite basin. Figure 7.2 shows the distribution of the individual turbidite systems in the stratigraphy. The names of the systems have been taken from Mutti *et al.* (1989), and refer to the localities where sandstone lobes and/or

their equivalent or slightly younger channel-fill deposits are best developed. Generally, each system is characterised by a clastic depositional system at the base, commonly associated with a phase of tectonic activity, succeeded by deposition of a largely mud-dominated overbank wedge.

Water depths in the Hecho Basin have been estimated by Mutti and Normark (1987) as being between 600 m and 2000 m. This depth range was based mainly on ichnofacies assemblages and is clearly subject to considerable uncertainty.

The following sections provide detailed descriptions of the facies and architecture of the channel, interchannel and feeder canyons for the Hecho Supergroup turbidite systems. For each system, emphasis is given to evaluating the channel/feeder system type from its architecture and facies, resulting in a chronological development of the feeder systems of the Hecho Supergroup. Aerial photographs of the area have allowed the large-scale mapping of the channel complexes of the Hecho Supergroup. Using the aerial photographs and structural measurements taken in the field, a stratigraphic restored section has been created (Figure 7.3).

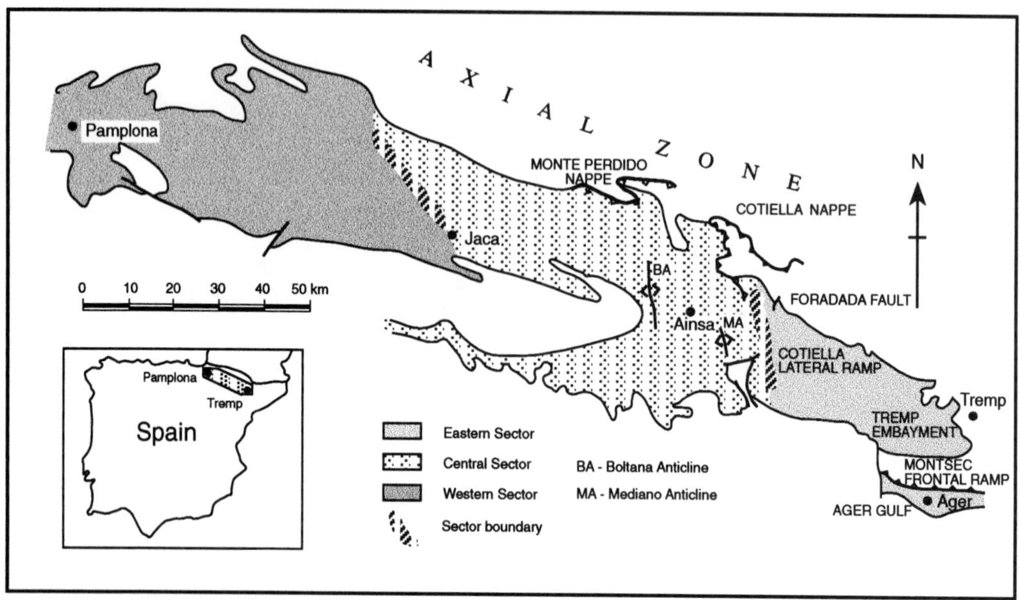

Figure 7.1. Map of the main palaeogeographic sectors of the Tremp-Pamplona Basin and the major structural elements (after Mutti *et al.* 1989).

W   E
Rio Ara   Ainsa Basin

UPPER CD	**GUASO SYSTEM**	GAUSO CHANNEL-FILL COMPLEX
	MORILLO SYSTEM	MORILLO CHANNEL-FILL COMPLEX
	AINSA SYSTEM	AINSA CHANNEL-FILL COMPLEX
LOWER CD	**BANSTON-FISCAL SYSTEM**	OVERBANK WEDGE
	FISCAL LOBES	BANASTON CHANNEL-FILL COMPLEX
SL	**GERBE-CONTEFABLO SYSTEM**	OVERBANK WEDGE
	CONTEFABLO LOBES	GERBE-CHARO CHANNEL-FILL COMPLEX
CS2	**BROTO SYSTEM**	OVERBANK WEDGE
	BROTO LOBES	LOBES & CHANNEL-FILLS
CS1	**FOSADO-TORLA SYSTEM**	OVERBANK WEDGE
	TORLA LOBES	FOSADO CHANNEL-FILL COMPLEX

Figure 7.2. Distribution of turbidite systems in the stratigraphy of the Tremp-Pamplona Basin, in the Río Ara and Ainsa areas (after Mutti *et al.* 1989). CS - Castisent Group; SL - Santa Liestra Group; CD - Campodarbe Group.

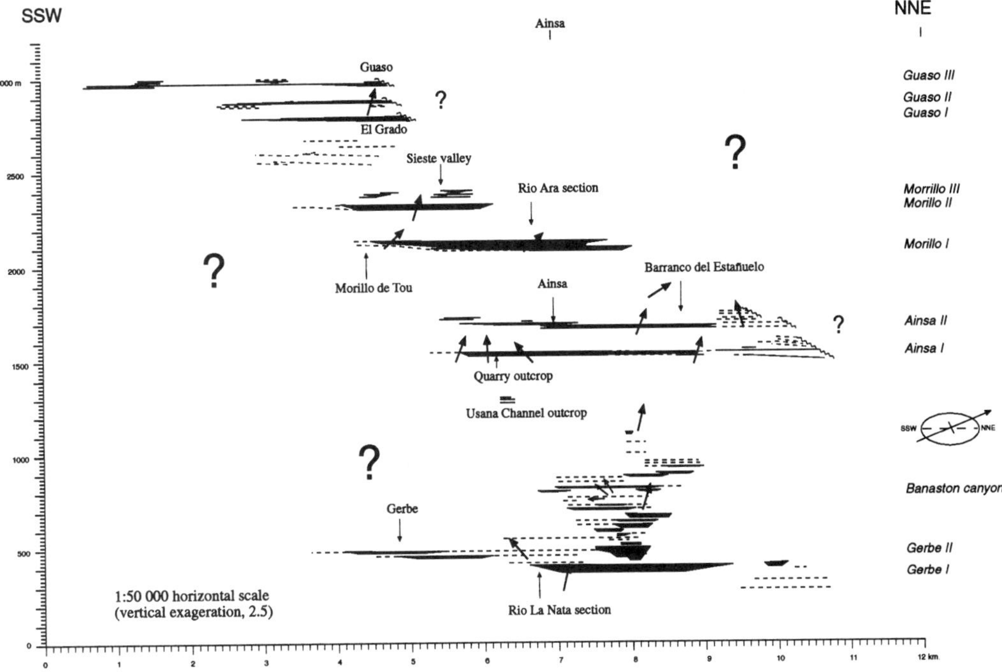

Figure 7.3. Restored section of the Ainsa Basin (based on aerial photograph interpretation) showing the vertical distribution of outcropping channel-levee complexes. Palaeocurrent arrows are orientated with respect to the orientation of the section (SSW-NNE), *i.e.*, vertical arrows indicate palaeocurrent directions towards the WNW.

Fosado-Torla System, Upper Castisent Group

Lower Castisent sediments (CS1) form a prograding sequence from the basin floor turbidite system, expressed by the Fosado Channel Complex, overlain by CS1 overbank wedge sediments, to shelf sandstones and mudstones.

Detailed descriptions of the architecture and facies of the Fosado Channel Complex deposits have not been included here because of their largely inaccessible exposures. Channel outcrops can be viewed across the Barranco de Fosado south of the village of Fosado. These channels belong to a Type I turbidite system (sensu Mutti & Normark 1987), and consist of thick sandy lobe deposits in the western part of the central sector near the village of Torla.

CS1 slope facies consist of mottled muds and silts, some muddy and silty sandstone beds, thin-bedded (1-3 cm thick) turbidites with rippled tops and a few isolated medium-/fine-grained, medium-bedded turbidites. Slumping and sliding are common in the CS1 slope facies in this area. Figure 7.4 shows a large-scale slide-scar and associated deposits within this succession. There is an increase in bed thickness immediately above the scar surface due to the ponding of turbidites within the submarine erosional feature.

Broto System, Upper Castisent Group (Charo-Arro System)

The Broto System in the vicinity of the town of Broto, forms a 1500 m thick succession of sandstone lobe/sheet-like deposits in the western part of the central sector, and continues a further 85 km westward at least as far as Roncal (Mutti *et al.* 1989). In the eastern part of the central sector, there is a significant feeder system for these deposits, represented by a unconformity into the Lower Castisent Group shelf and slope deposits, known as the Charo Canyon. A large siliciclastic deposit which has been termed the

Figure 7.4. Photograph of a large-scale slide structure in CS1 slope facies (figure for scale). The deposits of largely thin-bedded turbidites young towards the top left of the outcrop. Slightly thicker sandstone beds drape the infill lying above the erosion surface.

Arro Sandbody (Millington & Clark 1995) lies immediately basinwards from the Charo Canyon. The Arro Sandbody is stratigraphically equivalent to the Broto turbidite lobes, but does not show true channel features, unlike siliciclastic deposits of other Hecho Supergroup turbidite systems found in this palaeogeographical area. The lack of a major Upper Castisent channel-fill complex in the eastern part of the central sector Group, is the reason why the Broto System is not named after a channel complex and lobe equivalent system. In the following sections, the architecture and facies of the Charo Canyon and the Arro Sandbody are described and interpreted.

The overall coarser-grained nature of the fill of the Charo Canyon result in the canyon sediments being exposed on the Charo hill, surrounded by largely finer grained shelf and slope sediments. The canyon is incised into slope and shelf facies of the Castisent and Santa Liestra Groups, and contains complex fills of sediments belonging to these stratigraphic groups (Figure 7.5).

The Arro Sandbody, deposited at the base of the CS2 strata, consists of a ridge of outcrop, running from the Charo Canyon area to an area north of Ainsa, forming the Sierra de Arro and the Santa Catalina hill.

Figure 7.6 shows the stratigraphic relationships of the sediments of the Charo Canyon with the present day structural overprint removed (Sgavetti 1991). The Arro Sandbody and Charo Canyon have been mapped using aerial photographs, and from detailed fieldwork (Figure 7.7). The crucial outcrop relationships between the Charo Canyon, the incised shelf and slope sediments it incises, and the down-slope sediments of the Arro Sandbody, are somewhat obscured by structural complexities. The field relationship between the CS2 strata of the canyon-fill and the Arro Sandbody are complicated because: (i) the transition from canyon deposits to base of slope deposits occurs in the hinge of an overturned anticline; (ii) the anticline is cut by a minor thrust fault, and (iii) the outcrop of the Arro Sandbody dramatically pinches out in the vicinity of the Charo Canyon.

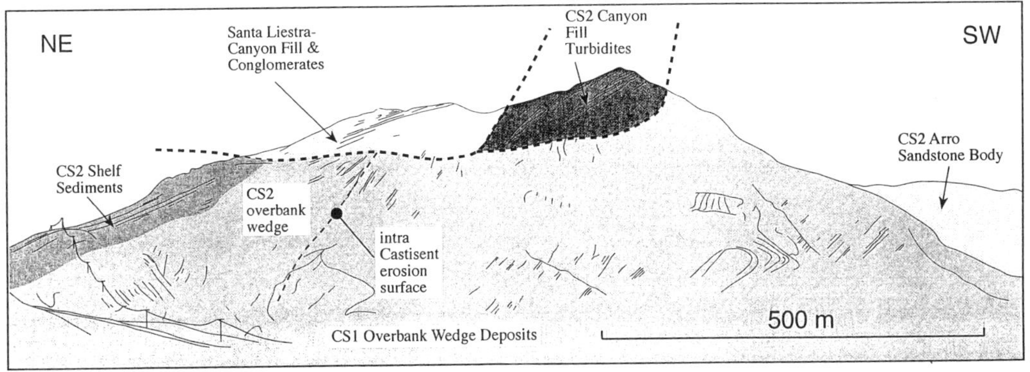

Figure 7.5. Sketch showing the major unconformities in the Charo Canyon. Upper Castisent (CS2) and Santa Liestra (SL) sediments make up the fill of the Charo Canyon incised into CS1 and CS2 strata (Millington & Clark 1996).

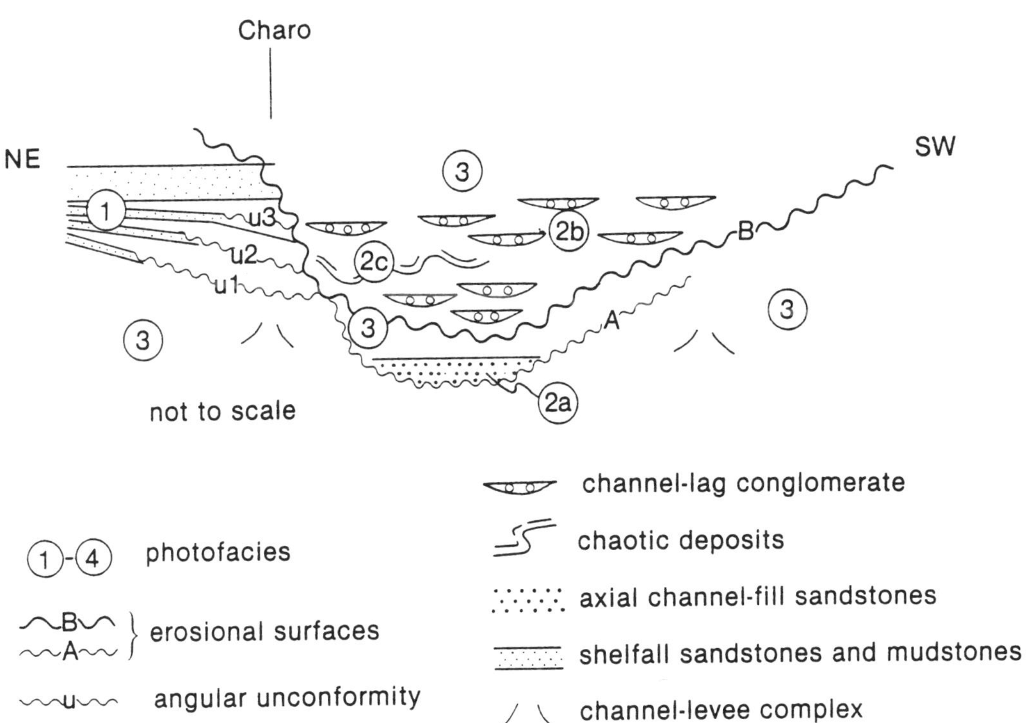

Figure 7.6. Restored section of the Charo Canyon fill showing the fill and stacking relationships of the Castisent and Santa Liestra strata, separated by unconformity B (after Sgavetti 1991). Photofacies are those of Sgavetti (1991). The canyon is approximately 1.5 km wide.

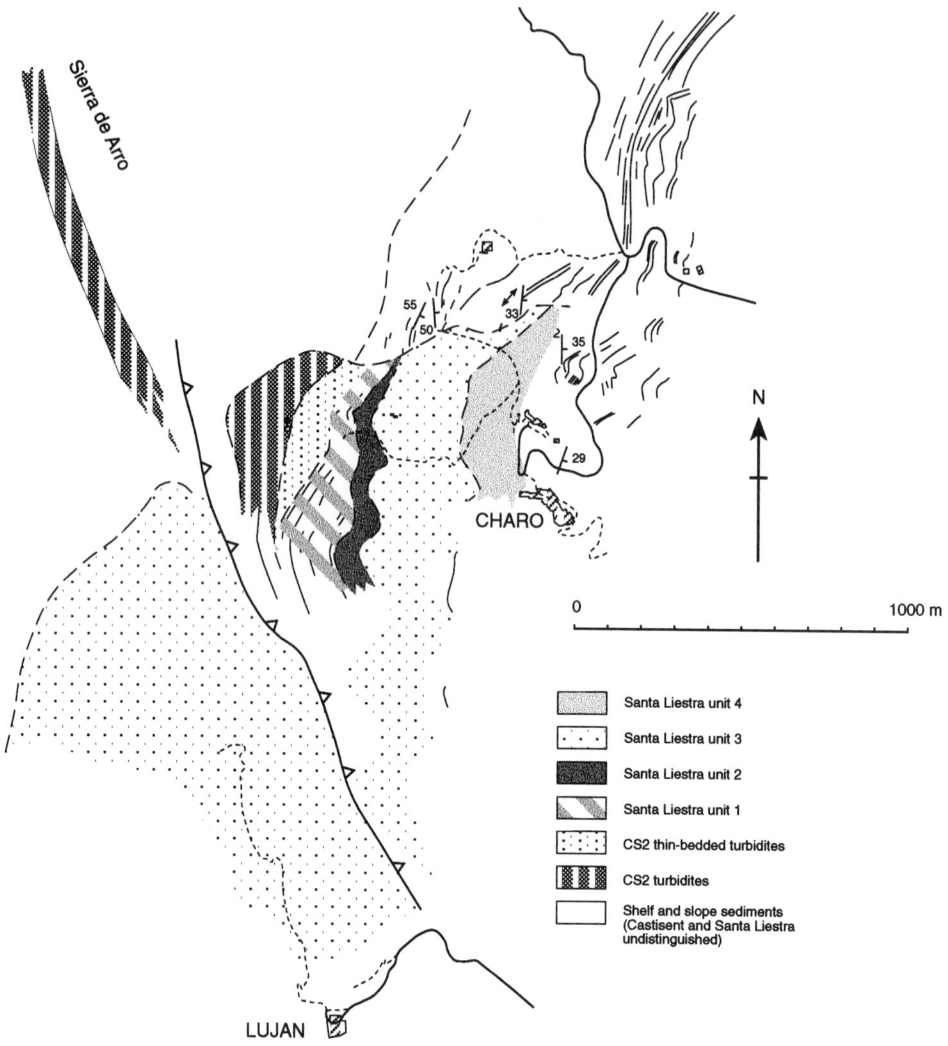

Figure 7.7. Map of the Charo Canyon area. After Millington & Clark (1996).

The Charo Canyon sediments are best exposed in the eastern limb of the Charo Anticline. This structure is a N-S trending S- plunging overturned anticline, containing a minor thrust fault running approximately through the fold axis. The larger-scale Mediano Anticline lies to the south-west of Charo, where Alveolina Limestone Group carbonates are found exposed in its core.

Above the CS1/CS2 unconformity on the western limb of the Charo Canyon, there are only a few turbidite beds reaching a total thickness less than 1 m. Above this unconformity on the eastern limb, 45 m of CS2 turbidites occur in the basal fill of the Charo Canyon. Uplift of the Mediano Anticline has resulted in the removal by erosion of the main axis of connectivity between the overlying Santa Liestra fill of the Charo Canyon, and the very coarse-grained fill of the time-equivalent Gerbe Channel. By taking into account the structural complexities of the area, sedimentological analysis of the deposits can be used to evaluate depositional relationships between shelf, slope, canyon and base-of-slope, depositional elements.

Charo Canyon sedimentary fill

Age	Eocene
Fan position	Canyon
Width	1500 m
Depth	400 m
Length	-
Aspect Ratio	3.75
Fill Type*	A
Max. Sinuosity	-

(see Figure 7.37)*

The basal fill of the Charo Canyon is incised into the CS1 slope and shelf deposits. Adjacent to the Charo Canyon itself, the CS1 slope deposits show many angular unconformities, resulting from post-depositional sliding and slide scars associated with collapse of the adjacent canyon walls.

Mass failures produced by tectonic activity and sedimentary over-steepening provided large amounts of sediment for the Broto Turbidite System. The Charo Canyon was subsequently infilled by the Broto System turbidites (CS2) (locally termed the Arro turbidites), and deep-marine mudstones of the Broto System overbank wedge. The Arro turbidites exposed at the base of the Charo Canyon consist of thick-, medium- and thin-bedded turbidites (max. bed thickness 2 m). The thick- and medium-bedded turbidites appear in "packets", and interpacket mudstone. Thin-bedded turbidite intervals are heavily bioturbated and show evidence of slide/creep deformation with dislocated and contorted bedding structures. Turbidites are commonly coarse- to medium-grained, graded beds, some having irregular bases with abundant flutes and groves. Also, there are a few conglomeratic beds. The sandstone:mudstone ratios gradually decrease in the upper part of this unit. Lateral pinching of beds is common. These turbidites are interpreted as being deposited from high-concentration, confined or channelised turbidity currents within the canyon. These flows carried large amounts of material which was subsequently deposited to form the Arro Sandbody and the associated thick Broto lobe sequences in the western part of the central sector. Above the channelised turbidite deposits, the CS2 fill of the canyon is represented by a succession of thin-bedded (1-

5 cm thick) turbidites with rippled tops. Measurements of some ripples show palaeocurrents 90°, divergent to the general depositional trend of the canyon. These directions are interpreted as side-wall reflected currents in the waning (final) stages of CS2 turbidite sand deposition.

CS2 shelf sediments flank the canyon area, and contain many relatively minor unconformity surfaces, indicating that tectonic activity continued throughout deposition of CS2 strata. These deposits consist largely of muds, containing abundant nummulites, carbonate nodules associated with red/orange marl horizons (1-2 cm thick), and some sandstones containing coral, bivalve, oyster and gastropod fragments, plant debris and nummulites. Sandstones commonly occur in "packets" between intervals of mainly mudstones and siltstones. These sandstones are slightly muddy, laminated, have rippled tops, and are commonly very bioturbated (horizontal and vertical burrows).

The Charo Canyon was reactivated as evidenced by the development of the Santa Liestra submarine erosion surface (Mutti *et al.* 1989). East of the Charo region the erosional surface shows evidence of large-scale sub-aerial erosion, and westward it is represented by the deposition of the deep-marine coarse-grained Charo-Gerbe channel-fill overlying CS2 overbank muds. The Santa Liestra fill of the Charo Canyon contains four separate mappable units, numbered 1-4 from the base. Unit 1 consists almost entirely of mudstone deposits with isolated medium-bedded turbidites (10-20 cm thick). These beds are relatively laterally continuous. Unit 2 contains distinctive facies of highly contorted mudstones, with numerous sandy-filled burrows, and muddy debrites with some "floating" cobble-sized boulders. Coarse-grained distorted and partially dismembered sandstone beds (30-50 cm thick) provide further evidence for post-depositional slope-related wet-sediment creep processes. Unit 3 is the thickest stratigraphic unit of the Santa Liestra canyon-fill, and consists of sandstone lens elements which show evidence of minor channelisation (Figure 7.8). These minor channel bodies are filled with medium- to very coarse-grained T_{abc} turbidites, and pebble-size conglomerates commonly containing numerous nummulites and mud clasts.

The channelised bodies typically consist of three to five beds, up to 2 m in thickness, suggesting that these channels were relatively short lived transient features. Associated with the sandstone lenses, and typically preceding their deposition, are minor scour elements (0.5-2 m deep, 1-5 m wide), filled with large boulders (up to 90 cm diameter), locally imbricated (Figure 7.9). The boulders and cobbles contain many oyster, bivalve and coral fragments, and limestone cobbles, commonly showing algal borings, indicating that some of the clasts may have been resident in a beach/near-shore environment preceding transport into the canyon. Between the channelised bodies of this unit are muddy debrites, and mudstone intervals which contain creep-deposited thin sandstones. The debrites also contain oyster fragment-rich boulders, a characteristic not observed in the boulders of older debrites. Unit 4 consists of thin- and medium-bedded, medium- to coarse-grained Tabc turbidites. Despite the relative high sandstone content of this unit, it tends to be poorly exposed throughout the canyon fill. This unit is interpreted as the final clastic fill of the Charo Canyon, with the Santa Liestra shelf and slope sediments overlying these deposits. The presence of turbidite deposits on top of the incised Castisent shelf deposits indicate dramatic rates of subsidence associated with the basal Santa Liestra unconformity (Mutti *et al.* 1989). The overlying Santa Liestra shelf deposits contain thin-bedded, medium- to coarse grained, nummulitic sandstones which are bioturbated. Some beds show symmetrical straight-sinuous crested wave-ripples. Interbedded mudstones contain carbonate nodules. Figure 7.10 shows a summary log of the Charo Canyon fill.

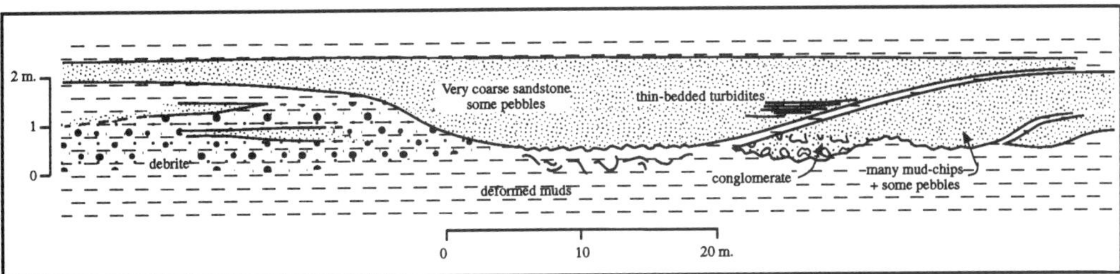

Figure 7.8. Sketch of a sandstone- and conglomerate-filled channel element within the Santa Liestra fill of the Charo Canyon After Millington & Clark (1996).

Figure 7.9. Photograph of minor scour elements of unit 3 filled with boulder-sized clasts.

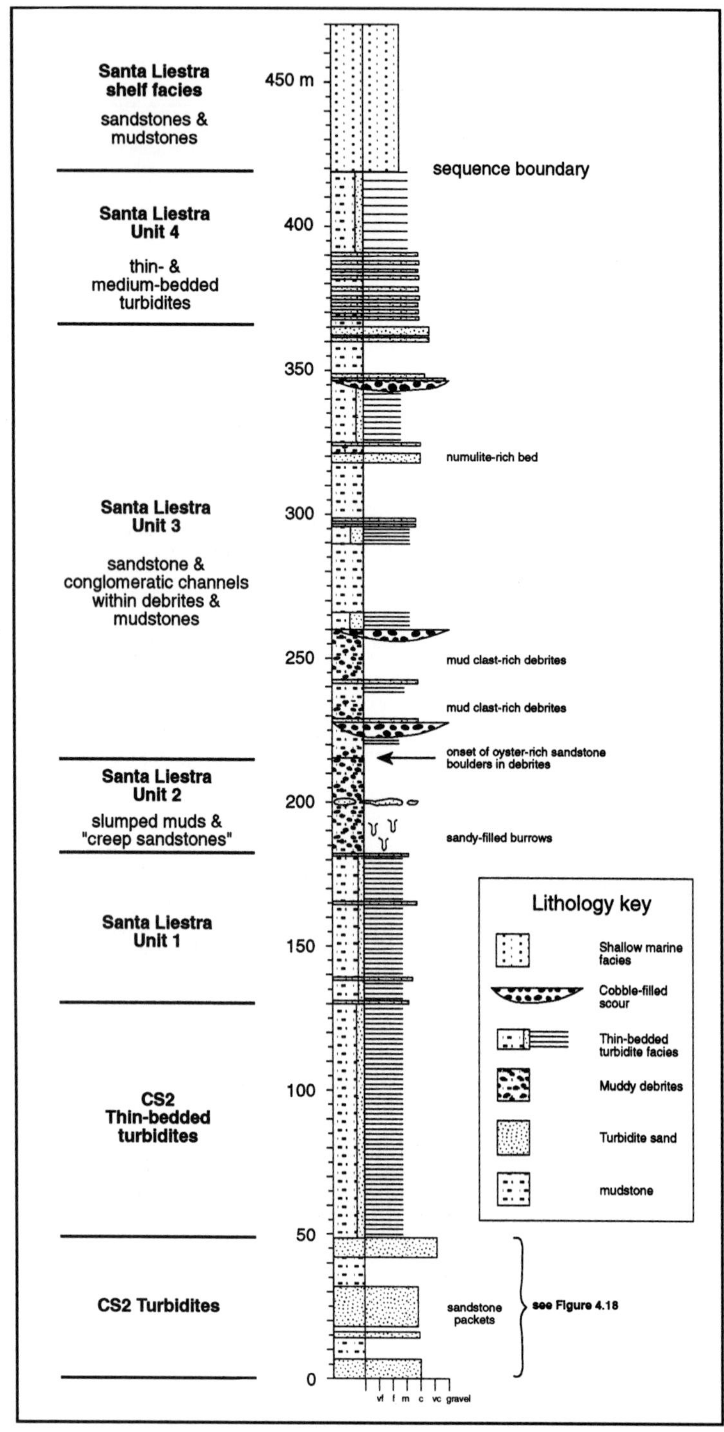

Figure 7.10. Schematic log of the Charo Canyon axial fill, from basal CS2 turbidites to Santa Liestra shelf facies. After Millington & Clark (1996).

Arro Sandbody

Age	Eocene
Fan position	Base of slope/canyon mouth
Width	>8000 m
Depth	140 m
Length	-
Aspect Ratio	>57
Fill Type*	G/C
Max. Sinuosity	-

(see Figure 7.37)*

The CS2 strata in the Charo region consists of the Arro Sandbody and associated overbank wedge deposits. Thrusting in the area has locally tilted the Sandbody near vertical and has even resulted in overturned strata in the Río Nata section. The Arro Sandbody is the only sandstone turbidite unit of the Hecho Supergroup that is continuously traceable into its canyon source, the Charo Canyon.

The Arro Sandbody is best exposed at the Río Nata section. Other exposures of the Arro Sandbody are described by Millington and Clark (1995), where detailed bed-by-bed sedimentary log sections have been measured along down-current localities illustrating the sedimentology of the Arro Sandbody.

The Río Nata section is approximately 140 m thick. The base of the section is dominated by coarse-grained erosive-based sandstone turbidites. Sediment draped scours indicate that this was a region of high flow turbulence and intense erosion (Figure 7.11). Many of the muddy intervals show characteristics of sediment slides. The section thins and fines upwards into a medium-bedded facies association, where beds are commonly in the order of 20-30 cm thick with erosive bases. Cross-bedding structures and out-sized mud clasts are typical of the erosive-based sandstones (Figure 7.12). The sandstones commonly show T_{bce}; the T_a division tends to be either absent or isolated without other Bouma sequence divisions. The tops of the beds of many of the medium-bedded sandstones display current ripples.

In the mid to upper parts of the section, muddy sediment slides predominate. Within the slides the muddy sediments show large open shear-fold structures. The topographic ridge in this area is a result of the thick structureless sandstone member which overlies the sediment slides. The structureless beds are very coarse-to coarse-grained sandstones, internally erosive, and show little if any lamination and contain numerous rip-up mud clasts. Above the thick structureless sandstone member, muddy sediment slides again dominate the section, with associated coarse-grained non-persistent erosive-based sandstones showing cross bedding and mud-draped scour surfaces. The top part of the section becomes increasingly mud dominated, interpreted to reflect the abandonment phase of the Sandbody deposition.

The mean palaeocurrent direction of the Río Nata section is approximately NNW which is more northerly than the palaeocurrents seen in the Charo Canyon CS2 turbidites, and those seen in more westerly outcrops of the Arro Sandbody (Millington & Clark 1995). This inflection of palaeoflow in the Río Nata section CS2 turbidites is believed to be attributed to localised topography in the Canyon mouth area, attributed to growth of the more southerly Mediano Anticline.

Palaeocurrents measured from ripples in the Arro Sandbody show a large divergence from the mean trend, which has been attributed to deposition from flow expansion processes as turbidity currents left the confines of the canyon (Millington & Clark 1995). Density currents undergoing flow expansion would initially record current orientations, represented by erosive flutes and grooves showing little palaeocurrent diversity. After flow expansion, turbidity current flow directions diverge and the flanks of the expansion zone would have a net deceleration. At any one locality the character of the turbidity current would change from being largely erosional to essentially depositional, with the ripples recording more diverse palaeocurrent directions. Ripple palaeocurrent data, therefore, is predicted to show a broad scatter with a mean constant similar to that of the groove and flute data, an observation made in the Charo canyon-mouth area.

Figure 7. 11. Photograph of an erosional scour-and-fill element within the overturned beds of the Arro Sandbody, Río Nata section. (beds young towards the right).

Figure 7.12. Photograph of cross-bedding structures within turbidite beds of the Arro Sandbody (Río Ara section).

Gerbe-Cotefablo System, Santa Liestra Group

The Santa Liestra Group of the central sector consists of shelf deposits, sediments deposited within the Charo Canyon, turbidite and debrite infill of the Gerbe Channel System, turbidite sandstone lobes (the Cotefablo lobes), and overbank wedge sediments. The sandstone beds of the Cotefablo lobes are relatively thin compared to those of the Broto System (Mutti *et al.* 1989). The Gerbe Channel Complex consists of two major channelised bodies termed the Gerbe I and the Gerbe II (see Figure 7.3).

Gerbe I Channel Complex

Age	Eocene
Fan position	Proximal fan channel
Width	2750 m
Depth	45 m
Length	-
Aspect Ratio	61
Fill Type*	B
Max. Sinuosity	-

(see Figure 7.37)*

The Gerbe I Channel outcrops conveniently along the Ainsa-Fuendecampo road (N140), and in outcrops exposed nearby in the Río Nata. At these localities the Gerbe I Channel can be divided into lower and upper channel-fill elements. The upper channel-fill element is dominated by coarse- and medium-grained turbidite sandstones, with the lower channel-fill element characterised by alternations of muddy and gravelly debrites.

The Gerbe I Channel is incised into laminated muds, silts (Facies D2.3) with rare thin and irregular fine-grained sandstone beds of the Castisent overbank sediments. The basal fill of the lower channel-fill element consists of an irregular coarse-grained turbidite (140 cm max. thickness), which was subsequently overlain by mainly debrites.

The lower channel-fill unit, dominated by the debrites, appears to consist of depositional cycles showing a sequence changing from mud-rich debrites, gravelly debrites, gravelly debrites with diffuse stratification, clast-supported conglomerates (some of which have thin, discontinuous coarse-grained sandstone horizons), to sandy turbidites. This succession may be partial or complete and repeated several times in the lower channel-fill. The mud-rich debrites (Facies A1.3) contain scarce, generally small (<5 cm) clasts, while the gravelly debrites (Facies A1.2) contain clasts up to 120 cm (Figure 7.13). Clasts are predominantly sandstone and limestone. Some debrites show evidence of scouring and subtle stratification, and associated with these deposits are clast-supported gravels (Facies A1.1, A2.1, A2.2). The few interbedded turbidite sandstones of this succession tend to be coarse-grained medium/thick-bedded, and contain dish-structures (Figure 7.14) and ripples (suggesting that deposition was initially from fluidised sediment flow or high concentration turbidity currents). The distinct facies associations and range of debrite types in the lower channel-fill element suggest that these deposits may have been formed from similar original flow types, which subsequently underwent progressive down-slope flow transformations. Lowe (1982) suggested that there is a continuous flow spectrum between cohesive debris flows and cohesionless liquefied flows, and that deposits from these flows would reflect this continuum, in the form of cohesive freezing deposits, frictional freezing deposits, and deposits from high and low concentration turbidity currents. Evidence for these processes occurring can be seen in the debrite and turbidite deposits of the lower channel-fill unit.

The architectural geometry of these lens-shaped deposits, result from either the irregular depositional topography of successively deposited debrites or down-channel scour-and-fill processes. As discussed earlier some of the "evolved" debrites show evidence for localised scouring.

The turbidite deposits of the upper fill of the Gerbe I Channel contain architectural elements and facies typical of channel-fill deposits (Facies Groups B2 and C2). The beds are typically discontinuous, medium-bedded Tabc, largely coarse-grained turbidites, and show an overall thinning-upward sequence.

Figure 7.13. A large (120 cm) rounded boulder within a debrite unit in the lower part of the Gerbe I Channel.

Figure 7.14. Coarse-grained turbidite with dish-structures in the lower part of the fill of the Gerbe I Channel.

Gerbe II Channel

Age	Eocene
Fan position	Proximal fan channel
Width	1500 m
Depth	30m
Length	-
Aspect Ratio	50
Fill Type*	D
Max. Sinuosity	-

(see Figure 7.37)*

The fill of the Gerbe II Channel is very different to that of the Gerbe I Channel. It consists of thick-, medium- and thin-bedded Tabc and Tbcd sandstones (almost entirely Facies C2.1, C2.2, C2.3). There are no conglomeratic beds, although some beds contain abundant out-sized mudclasts at their bases or along amalgamation surfaces within beds. Figure 7.15 shows a 15-m-thick sequence of channel-fill sandstones (maximum bed thickness is 210 cm). This sand-rich sequence represents the axial fill of the Gerbe II Channel, and is overlain by another 15 m succession of mudstones, thin- and medium-bedded sandstone turbidites. The sandstones of the mud-rich succession appear in three thinning-upward packets, interpreted as representing the final infill (generally passive) of the Gerbe II Channel.

The Gerbe II Channel exhibits spectacular bed pinch-out from the channel axis towards its margins. Over a lateral distance of 100 m, the sandstone beds of the channel-fill lens out entirely into surrounding overbank mudstones. Figure 7.16 shows the pinch-out of these beds at the channel margin. Palaeocurrents from ripples in the channel margin deposits record a WSW orientation, while the channel-axis deposits record palaeocurrents (from sole structures and ripples) towards WNW. The approximate 45 degrees deflection of palaeocurrents at the channel margin is interpreted to have resulted from turbidity currents riding up the channel margin, and flowing back towards the channel axis.

Figure 7.15. Thick-bedded turbidites of the axial-fill of the Gerbe II Channel, viewed from Gerbe village.

Figure 7.16. Photograph of the Gerbe II Channel showing lateral pinch-out of channel-margin sandstones to the right. The section is approximately perpendicular to palaeocurrent, and approximately 100 m south (left) from this locality, thick-bedded turbidites of the axial channel-fill can be found (see Figure 7.15).

Along strike from the channel margin, a unit of thin-bedded turbidites developed (Facies C2.3, D1, D2.3), interpreted as channel levee facies. These deposits show relatively large-scale, low-angle surfaces (interpreted as slide scars), with associated overlying draped sediments (Figure 7.17). These beds can be traced a further 2.5 km north-east and perpendicular to palaeocurrent, where the unit reaches a thickness of 24 m, and is associated with a progressive increase in bed thickness. Sandstone beds are medium-grained thin- and medium-bedded T_{cd} turbidites interbedded with mudstones (sandstone/shale ratio of 50:50) (Figure 7.18). Slide surfaces and draped fill deposits are rare, and only seen on a small scale. Palaeocurrents are generally towards the WNW, parallel to those of the main channel axis. Due to the increases bed thickness, minor sedimentary deformation, and channel-axis-parallel palaeocurrents, these deposits are interpreted as ponded interchannel deposits (Figure 7.19).

The levee and ponded interchannel deposits form a stratigraphic horizon which can be identified on aerial photographs, showing its changes in thickness from channel to channel-margin, levee and interchannel deposits. A further 500 m north-east, along strike from the previous outcrop, a thick succession of sandstones develops (at least 50-60 m in thickness). Despite the high sandstone content of these deposits, they are generally poorly exposed and can only be correlated with neighbouring exposures by using aerial photographs. These facies consist of thick-bedded (1-2 m), coarse- to very coarse-grained, poorly sorted sandstones, with abundant mud clasts. From the aerial photographs, these deposits show a localised thickening in this area, with no evidence for channel facies. These observations lead to the interpretation that these sediments represent a further ponding of the Gerbe II overbank sediments, perhaps into a very localised submarine depression. Since this outcrop lies directly below the Banaston Canyon Complex (a sequence of preferentially vertically stacked channels, see next section), it is quite possible that the reason for the location of this thick sediment deposit in this area, may be due to control from an unexposed syn-sedimentary submarine fault, which did not necessary breach the seabed, but controlled the persistence of a localised depression on the seafloor. The localised ponding of these interchannel deposits, represent the initial development of the stacked channel sequence of the Banaston Canyon Complex.

Figure 7.31. Thin-bedded turbidite facies, interpreted as the Ainsa I Channel margin/levee deposits, on a road cutting 1 km north of Ainsa. The mud-rich deposits overlying the thin-bedded sandstone packet, contains large-scale slide folds.

Figure 7.32. Photograph of the thin-bedded turbidite facies of Figure 7.31.

Ainsa II Channel Complex

Age	Eocene
Fan position	Inner fan
Width	1500 m (channel complex)
Depth	50 m
Length	-
Aspect Ratio	30
Fill Type*	D
Max. Sinuosity	-

(see Figure 7.37)*

The Ainsa II Channel Complex can be viewed across the Río Cinca (Figure 7.33). The exposure forms a north-south trending (slightly oblique to palaeocurrent direction), 1.5 km cliff face up to 50 m in height. Down-cutting surfaces and most of the thicker beds can be "followed out" with the aid of binoculars from this viewpoint. Five stacked channelised bodies are identified, and numbered 1-5. The scale of the exposure enables comparison with the best resolution seismic records of stacked channels.

Abseil techniques enabled sedimentary logging of the channel complex, to establish precise bed correlations and document detailed aspects of the sedimentary facies comprising the channel fill. Figure 7.34 shows the detailed log correlation of the Ainsa II Channel Complex.

The sedimentary logs show that the channel fill sequences are dominated by turbiditic deposition. The most common facies present are parallel stratified sandstones, typically amalgamated, and amalgamated graded sandstone units showing Bouma Tabc divisions (facies B2.1 & C2.1, C2.2). Also present, and constituting a significant proportion of the channel facies, there are thick- and medium-bedded disorganised sandstones (facies B1.1) showing large scale dewatering structures. Relatively clean cross-stratified sandstones (facies B2.2) are uncommon throughout the channel fill complex, but tend to be found only in the basal beds of channels no. 1 and 2, and locally in the intermediate stage of fill of channel no. 4. This facies is interpreted to represent sediment bypass, and with respect to channels no. 1 and 2, suggests that an initial phase of sediment bypass existed prior to a much more depositional phase of sedimentation. Intercalated between these deposits, there are a few debrites, muddy sandstones (facies C1.1) and, less commonly, disorganised gravelly mud (facies A1.3). Mid-way up the fill of channel no. 4, one such muddy debrite can be correlated across the channel, although generally these debrites are not seen as being laterally persistent, implying either very localised or confined deposition from these flows. Thin-bedded turbidites (facies C2.3) and silty turbidites (facies D2.2) separate the packets of mainly sand deposition. These were deposited during relatively quiescent periods between the minor phases of sandy turbidite deposition (e.g. in channel no. 2 and 3), or away from the main channel axis depocentre to form the channel margin/levee facies (e.g. the thin-bedded turbidites, silts and muds flanking the small-scale channelised sandstone deposits in channel no. 3). Structureless muds (facies E1.1) occur in minor amounts within the channel, typically poorly exposed, and occur below the channel complex (seen at the base of logs 8, 11 and 14).

Palaeocurrent direction is generally towards the NW-WNW with a slight trend in modal current direction from NW to WNW with time. Some measured palaeocurrent directions show anomalous trends. Log 14 shows a palaeocurrent direction recorded perpendicular to the general palaeocurrent trend, near the margin of channel no. 2. This flow may have resulted from currents generated on levees, descending from the channel margin towards the channel axis. Channel no. 5 shows palaeocurrents ranging from SSE to WNW. The large variance of these palaeocurrent directions is believed to be a result of large-scale synsedimentary bank collapse and slumping at the channel margin. The steep margin of this channel, seen on the photograph (see Figure 7.33) to be very arcuate, may be considered as a slide or slump scar, rather than a down-cutting channel unconformity. Other evidence to

Channel-fill models for the Eocene Hecho Group submarine channels

Channel-fill Type		Examples
A 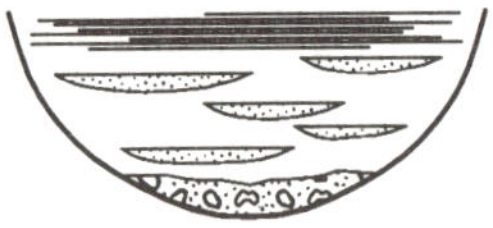	Stacked channel fill complex infilling large submarine erosional feature	Charo Banaston Guaso II Guaso III
B 	Erosional channel, most sediment bypasses channel area	Gerbe I
C 	Mixed erosional/depositional channel, represented by an initial phase of erosional channel characteristics, folowed by a depositional phase	Ainsa I Morillo II
D 	Depositional sand-filled channel	Ainsa II (Ainsa) Gerbe II
E 	Depositional channel, filled with alternations of sand and mud	Morillo III
F 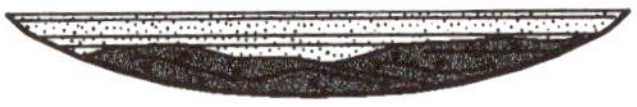	Submarine channel associated with large slide events. The base of the channel contains large ammounts of muddy debrites	Morillo I Usana
G 	Confined-unconfined flow transition deposits, containing abundant scours, cross-bedding and small-scale channel features	Arro Guaso I Ainsa II (Boltaña)

Figure 7.37. Summary of the channel-fill models for the Eocene Hecho Group submarine channels.

age	Castisent (CS2)		Santa Liestra		Lower Campodarbe		Upper Campodarbe									
channel complex	Charo Canyon	Arro Sandbody	Gerbe I	Gerbe II	Banaston Canyon	Usana	Ainsa I	Ainsa II		Morillo I		Morillo II	Morillo III	Guaso I	Guaso II	Guaso III
location	Charo	Rîo Nata	Gerbe	Gerbe	Banaston	Usana	Ainsa	Ainsa	Boltaña	Morillo	Rîo Sieste	Rîo Sieste	Rîo Sieste	El Grado	Guaso	Guaso
scale (of the coarse-clastic fill of the channels) — width (m)	1500	>8000	2750	1500	2000	(50)	(2000-3000)	750 (individual chs.), 1500 (ch. complex)	-	500	3000	3000	(100) (individual chs.), 2000 (ch. complex)	-	200-2000 (individual chs.), 2500 (ch. complex)	200-2000 (individual chs.), 2500 (ch. complex)
depth (m)	400	140	45	30	550	15	40	40 (individual chs.), 50 (ch. complex)	55	20	75	(35)	(5-10) (individual chs.), 50 (ch. complex)	70	5-10 (individual chs.), 150 (ch. complex)	5-10 (individual chs.), 150 (ch. complex)
aspect ratio (w:d)	3.75	>57	61	50	3.6	(3.3)	(50-75)	18.75, 30	-	25	40	(85)	(10-20), 40	-	(40-200), 17	(40-200), 17
channel type	A	G	B	D	A	F	C	D	G	F		C	E	G	A	A
Channel-fill — % sand	20%	55%	20%	50%	5-10%	65%	70%	80%	85%	50%	50%	65%	30%	85%	15%	15%
% mud	55%	20%	30%	50%	~60%	10%	10%	15%	10%	40%	25%	10%	60%	15%	50%	50%
% debrite	25%	25%	50%	0%	~30%	25%	20%	5%	5%	10%	25%	25%	10%	0%	35%	35%
associated lobe deposits	BROTO LOBES	BROTO LOBES	COTEFABLO LOBES	COTEFABLO LOBES	FISCAL LOBES	?	◄— (ponded sediment in front of Boltaña Anticline) —►									

Table 7.1. Summary table of the characteristics of different submarine channels and canyons in the Tremp-Pamplona Basin. Figures in brackets indicate uncertain values, and where necessary, the characteristics of the individual channel-fill elements (individual chs.) have been distinguished from the channel-fill complex as a whole (ch. complex).

Castisent to Upper Campodarbe ages. Approximations are given for the dimensions of the coarse clastic fill of each channel, which, despite the likelihood that most channels are finally infilled with mudstones and siltstones, are assumed to represent channel width and depth measurements. From these measurements, the aspect ratios of the submarine conduits clearly highlight the similarities between the Charo and Banaston canyons, and their differences from most of the other channel complexes in the basin with higher width:depth ratios.

Figure 7.3 shows that there is an apparent overall southward migration in the position of the channel axis. This shift is consistent with the largely southward migration of the stacked channel sequence of the Ainsa II Channel Complex. Because of the apparent consistent channel-axis offset direction, it is likely that most stages of turbidite system development were initiated by pulses of tectonic activity accompanied by the southward migration of the thrust stack to force the relocation of the basin axis southwards also: this contrasts with the classic foreland-basin model, where the migration of a peripheral bulge is predicted as towards, rather than away from, the thrust load (*cf.* Quinlan & Beaumont 1984).

Turbidite systems which contain more than one sand-rich sequence, commonly show a tendency to develop channel sequences which show deposition from the more erosional type channels and evolve to more depositional type channels (e.g. Gerbe I and Gerbe II channels, Ainsa I and Ainsa II channels, and the Morillo Turbidite System). In these cases, channel development may be the result of processes which do not include direct tectonically-induced variations in depositional style and rate and, or, eustacy, but rather periods of basin-fill and slope progradation between major changes in sedimentation induced by tectonics or eustacy. This latter scenario represents an alternative explanation to that offered in the preceding paragraph. The available data does not permit a unique solution to the problem posed by the consistent southward migration of the Ainsa II channel-axis sedimentation.

The margins of some of the channels can be roughly mapped-out across the Ainsa Basin over distances of the order of 8-15 km showing the approximate planform geometry of these channel systems. For the channels where this could be done, they all showed a

relatively straight channel course, with a slight northward bend in channel direction adjacent to the Boltaña Anticline. These results are consistent with the measured Palaeocurrent data and suggest that the Boltaña Anticline was an emerging palaeotopographic ridge influencing channel course during the fill of the Tremp-Pamplona Basin.

a

b

San Clemente nested channels, Capistrano Formation, California

Age	Late Miocene
Fan position	Middle fan
Width	750 m
Depth	20 m
Length	-
Aspect Ratio	37.5
Fill Type*	C/D/E
Max. Sinuosity	-

(see Figure 7.37)*

The Late Miocene deposits exposed in cliffs at San Clemente, California are assigned to the lower part of the Capistrano Formation. The sections show one of the best documented examples of the sequential lateral shifting of channel deposits (Figure 7.38). The eight nested submarine channels described and numbered by Walker (1975c) are bound by third-order channel erosion-surfaces (dividing groupings of second-order complexes), and are infilled with turbidite sands, pebbly sands, and muds. The stacked channels are interpreted as laterally avulsing middle-fan channels, which show a progressive southward swing in palaeocurrent direction (Walker 1975c) (Figure 7.39). The bases of channels numbered 2 to 8 are defined by mudstone drapes which Walker suggests are deposited from the upper (finer-grained) parts of turbidity currents that flowed through channel axes further incised into the seafloor sediments, i.e. the mud drapes were deposited relatively high-up on the channel margins. Subsequent flows gradually infilled the channels, so that coarser-grained material onlaps older mud drapes. Evidence for this process occurring can be seen in the fill of Channel #2, where sandstone beds can be seen to pass laterally into an amalgamated silt and mud unit at the channel base.

Re-examination of these deposits using the architectural element scheme have led to additional interpretations of these channelised deposits. Third-order bounding surfaces can be seen within the fills of Channels #3 and #8 (see Figure 7.39). These surfaces, by definition, have the same importance as the surfaces defining the bases of the numbered channels.

The sedimentary fill of each third-order channel element may be characterised by smaller scale observations, for example, Channel #2 contains a series of second-order sheet sand elements bound by second-order bounding surfaces (i.e. surfaces that delineate deposits formed from similar flow processes). The stepped-like erosion feature above the fill of Channel #2 can be defined as a third-order bounding surface due to its dramatic erosion of underlying second-order architectural elements. It is likely therefore that this represents the base of the third channel. Channel #3 contains additional third-order bounding surfaces (e.g. the stepped-cutdown surface and mud-drapes seen in Figure 7.39) which are therefore interpreted as representing channel-margins themselves. Channel #8 also contains a third-order bounding surface.

The northern cliff section at San Clemente shows a different assemblage of channel facies. There are no mud drapes, the section has a sand:mud ratio of nearly 100%. There is a high degree of amalgamation, and inclined channel margins (third-order bounding surfaces) dip predominately southwards. These deposits are interpreted as the channel axis facies of the northwards-migrating stacked channel margins seen in the cliff section south of the car-park (see Figure 7.39). This interpretation allows the half-width of Channel #8 to be measured suggesting a channel width of approximately 750 m.

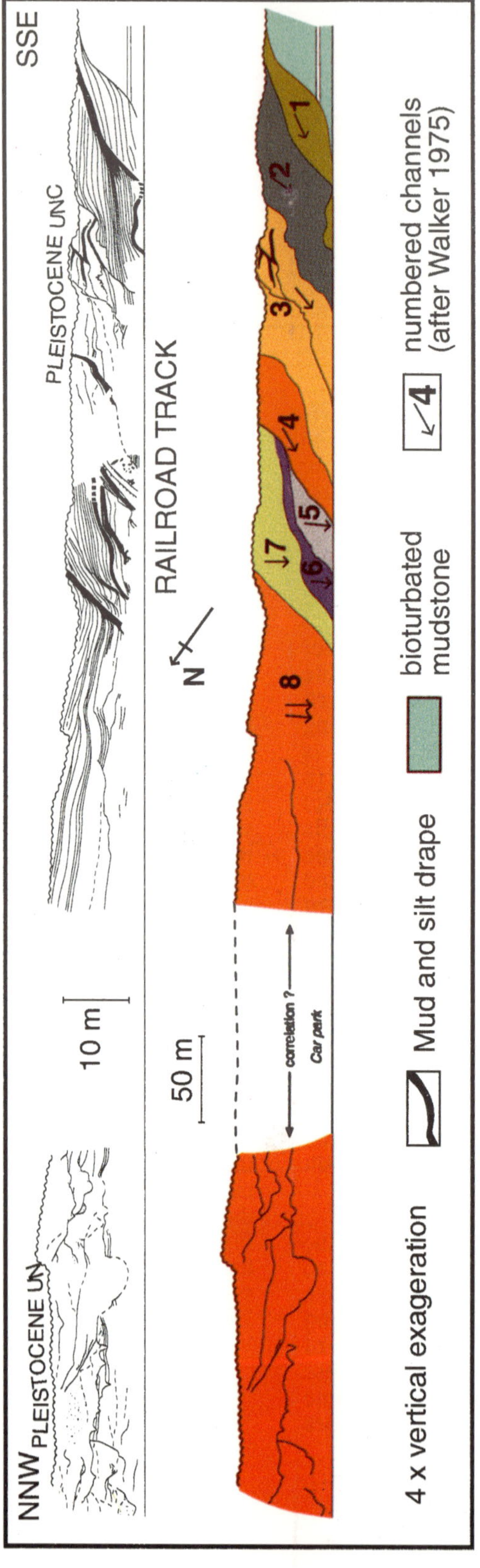

Figure 7.39. Eight nested channels in the Capistrano Formation at San Clemente, California, USA.

Upper Cretaceous Rosario Group, La Jolla, California

Age	Late Cretaceous
Fan position	Inner fan / canyon
Width	15 - 20 m
Depth	5 m (average)
Length	-
Aspect Ratio	3 - 4
Fill Type*	C
Max. Sinuosity	-

(see Figure 7.37)*

The Upper Cretaceous marine sediments of the Rosario Group accumulated in a forearc basin, contain a variety of facies associations, including those interpreted as middle- and inner- fan channels (Nilsen & Abbott 1981) (Figure 7.40). These channel facies occur in the Point Loma Formation and the Cabrillo Formation exposed at La Jolla and Point Loma, west of San Diego, California (Figure 7.41). The base of the Cabrillo Formation is defined by the first appearances of conglomerate in the marine deposits. The submarine fan built up on the landward-slope of the Upper Cretaceous forearc basin from westward flowing turbidity currents.

Nested inner-fan channel and canyon sediments of the Cabrillo are exposed in the cliffs at Tourmaline Surfing Park (La Jolla). The sections predominantly consist of thick-bedded sandstones and conglomerate lenses, with some medium-bedded sandstone facies. Architectural elements have a high degree of interconnectivity caused by the abundant amalgamation, scouring, and nesting of channel elements in this section (Figure 7.42). Large-scale sedimentary slide elements are also apparent in the Cabrillo Formation at Tourmaline Surfing Park (Figure 7.43). The channel elements contain disorganised and amalgamated architectural elements, indicating rapid cut and fill of these features. Nilsen and Abbot (1981) suggested that during the deposition of the Cabrillo Formation, sediment load from fluvial streams was deposited in deep-marine canyons incised very near to the coast-line.

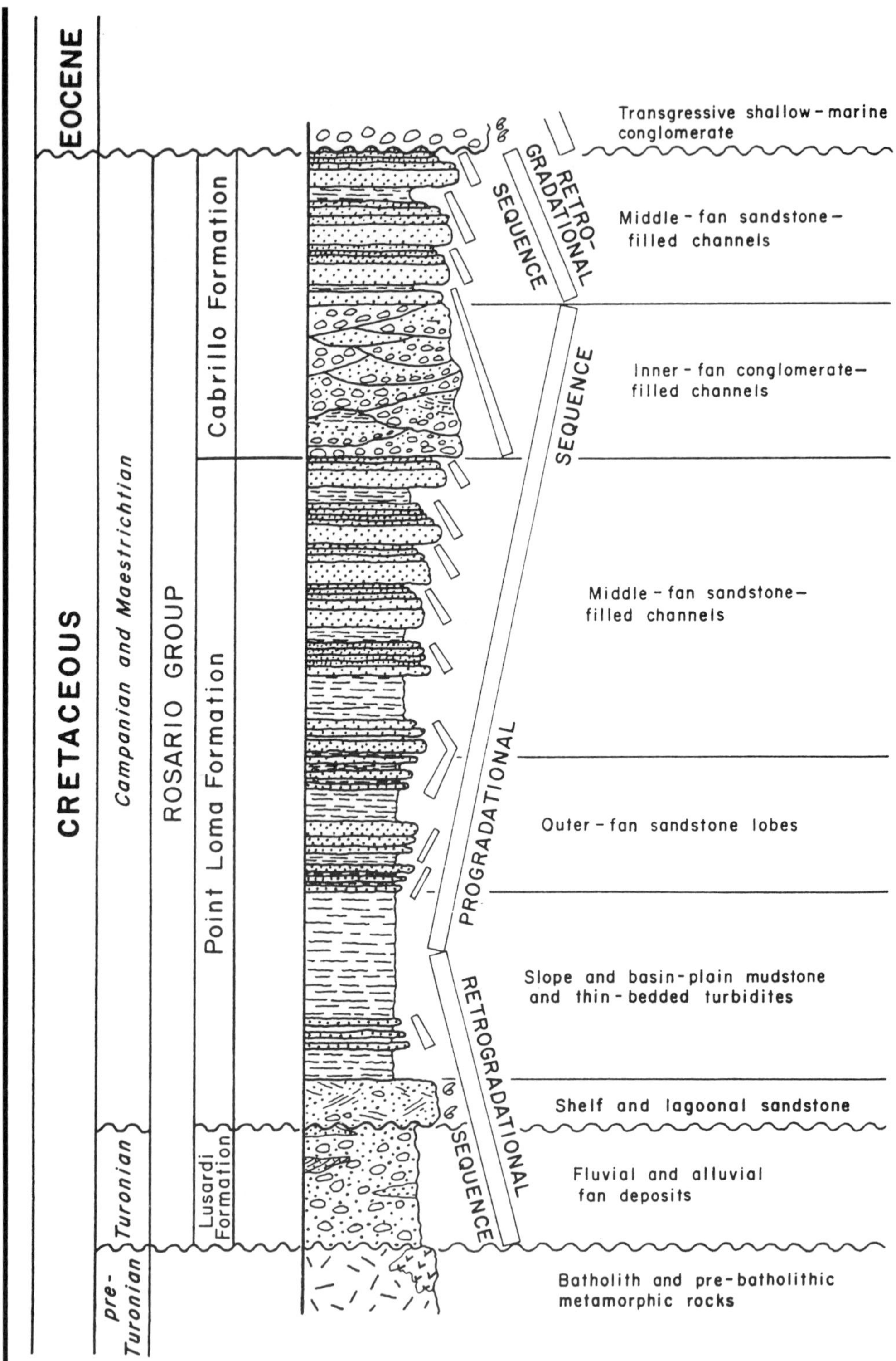

Figure 7.40. Stratigraphy and palaeoenvironments of the Rosario Group, California, USA (Nilsen & Abbott 1981).

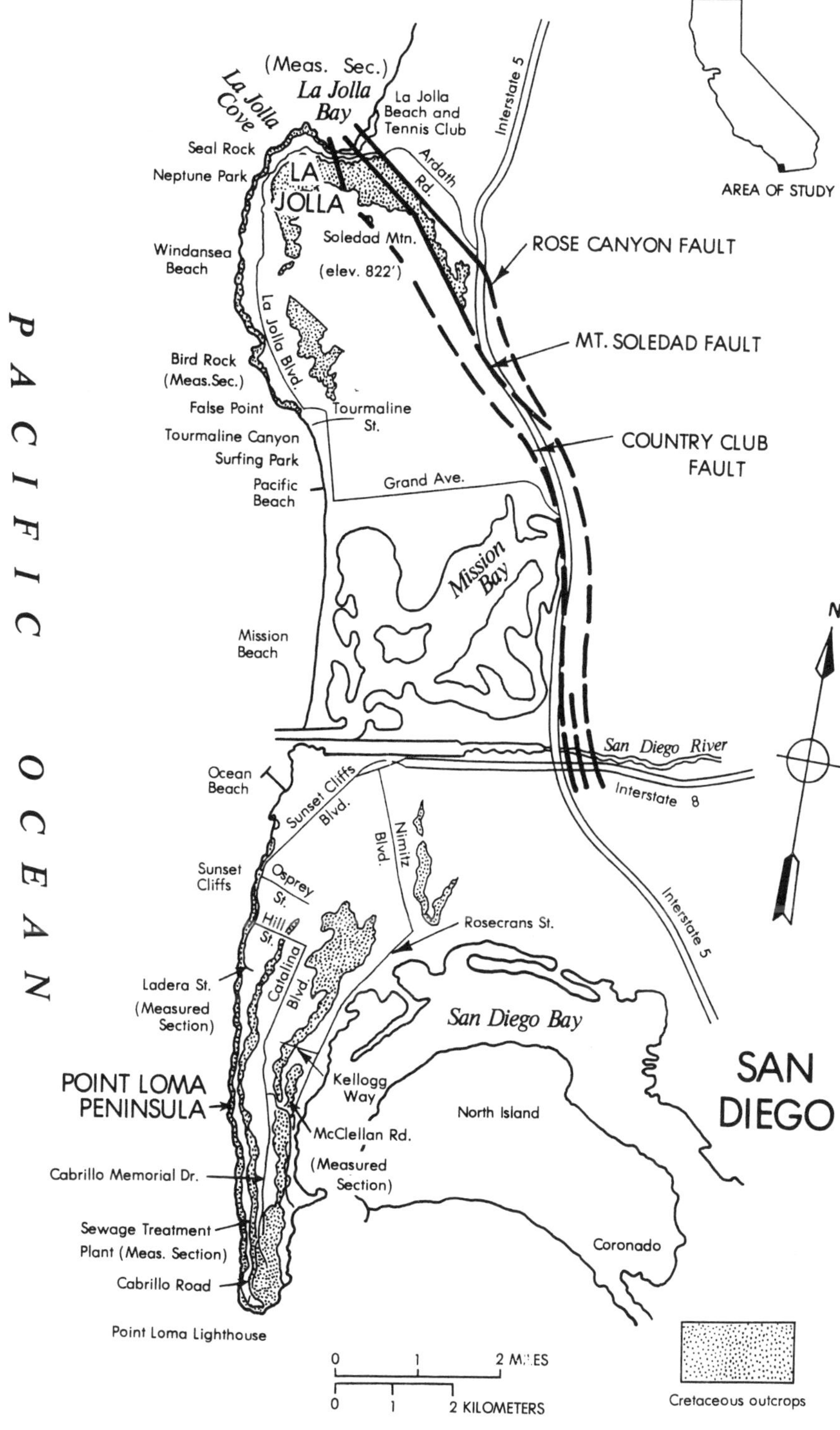

Figure 7.41. Location of Rosario Group outcrops in the San Diego area (Nilsen & Abbott 1981).

Figure 7.42. Conglomerate-filled scour (center), incised by a sandstone scour-and-fill element. There is an abundance of other small-scale scour-and-fill elements in this section (Cabrillo Formation, Tourmaline Surfing Park).

Figure 7.43. Large-scale sedimentary slide element in the Cabrillo Formation at Tourmaline surfing Park. To the left (north) the slide has a listric shape, and convoluted strata can be seen to the right (south).

Upper Cretaceous Chatsworth Formation, Simi Hills, California

Age	Upper Cretaceous
Fan position	Middle fan
Width	1000 m
Depth	15 m (average); 60 m (max.)
Length	-
Aspect Ratio	17 - 67
Fill Type*	D
Max. Sinuosity	-

(see Figure 7.37)*

The Upper Cretaceous Chatsworth Formation is exposed in the Simi Hills in the north-west suburbs of Los Angeles. In this area the Chatsworth Formation provides subsurface oil reservoirs. The Chatsworth Formation is part of a sand-rich deep-sea fan that possibly includes similar-aged deposits of the Santa Monica and Santa Ana Mountains (Link *et al.* 1984).

Channels are exposed on vegetated hillsides in the area, allowing large-scale approximations of channel aspect ratio, net:gross ratios and channel stacking patterns to be evaluated (see Figure 7.44). Channel palaeocurrents indicate a northwards flow direction, whereas interchannel sediments record more divergent flow directions between north-west and north-east. The channel-fills commonly comprise of thick-bedded sandstones which may be structureless or contain large-scale cross-beds or water escape structures. Detailed examinations of channel-fill architecture can be observed in the Simi Valley Freeway cutting which shows interfingering of channel and interchannel deposits (Figure 7.45). The channels are interpreted as forming in the middle-fan environment (Link *et al.* 1984).

Figure 7.44. View of the stacked depositional middle-fan channels in the Chatsworth Formation, Simi Hills, California.

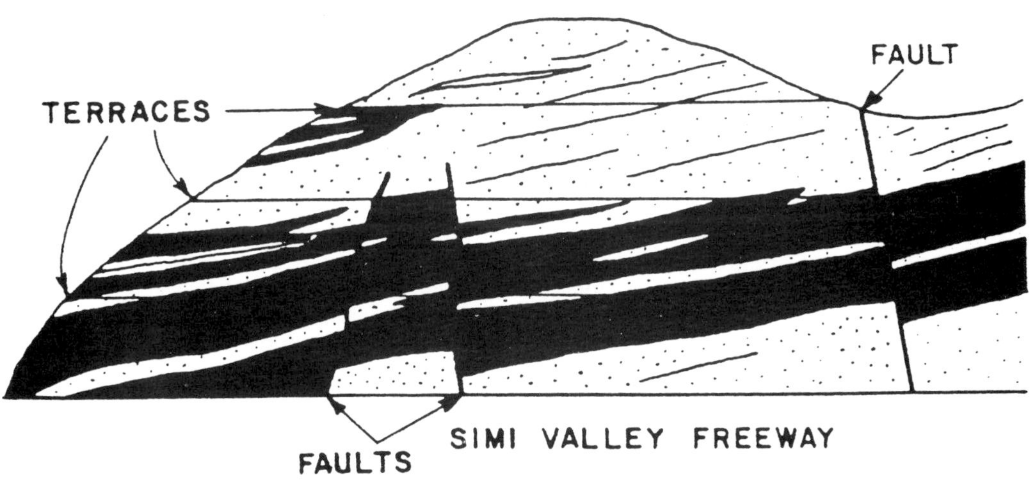

Figure 7.45. Sketch of the interfingering channel and interchannel deposits shown in a road-cutting, Simi Valley Freeway (Link & Squires 1981).

Miocene (Tortonian) Solitary Channel, Tabernas Basin, SE Spain

Age	Miocene
Fan position	Inner fan
Width	200 m
Depth	40 m (average)
Length	8 km (minimum)
Aspect Ratio	5
Fill Type*	C/F
Max. Sinuosity	-

(see Figure 7.37)*

The Miocene Solitary Channel is part of the Tortonian deep-sea fan deposits of the Tabernas Basin, SE Spain. Work by Kleverlaan (1987, 1989) has resulted in the delineation of three turbidite systems; System I (Sandy System), System II (Mixed System), and System III (Solitary Channel) (Figure 7.46). The trend of the Solitary Channel can be mapped for at least 8 km, although Cronin (1995) suggests that it can be mapped for 11 km. A relatively high width:depth ratio (5) suggests that the Solitary Channel may be a submarine canyon (a view also expressed to us by P. Haughton *pers. comm.*) (compare the aspect ratio with those of the Charo Canyon and Banaston canyon in the Hecho Group).

Due to the coarse-grained nature of the channel-fill facies, there is a lack of uni-directional palaeocurrent indicators, such as flutes and ripples. Recent field studies have identified imbrication, ripple, flute and over-turned flame structures indicating flow towards east-northeast at the western end of the channel. No unique flow directions were obtained from outcrops south of Tabernas village, i.e. it still remains unclear if these outcrops are a continuation of the Solitary Channel that outcrops west of the Granada-Almeria road.

The mapped outcrops of the Solitary channel follow the course of a major fault complex from the west to the Granada-Almeria road (Figure 7.47). The fault complex shows both normal and strike-slip faulting, and some of the fault spays show evidence of being growth faults during the deposition of the channel sediments. This implies that the course of the Solitary channel was very probably controlled by seafloor topographic expressions of an active fault. Many of the channel margin deposits are displaced either by complex later-stage reactivations on the channel-course controlling fault, or are highly tectonised due to differential compaction faulting in channel margin deposits where there is less coarse-grained material (Figure 7.48). Where preserved channel margins are observed, they show relatively steep onlap surfaces (Figure 7.49). Differential compaction related faulting can be expected to be an important feature of many subsurface channel bodies, and more detailed models may be useful in predicting the displacement of reservoir sands at channel margins.

The channel is filled with sandy debrites (Figure 7.50), conglomerates and some coarse-grained turbidites. In one of the more proximal sections, very large, angular and poorly sorted clasts constitute the fill of a large-scale scour element (Figure 7.51). The scour has a high aspect ratio, and may be a thalweg or a large channel-floor flute that has later been plugged by deposits from a debris flow.

Bedforms and inclined architectural elements are common within the channel-fill, perhaps due to the flow-freezing of channelised debris flows (Figure 7.52). Second-order bounding surfaces delineate lateral accretion clinoforms, and represent reactivation surfaces formed from changes within the channel flow, such as thalweg migration, increased sedimentary load, or tectonic and eustatic variations in the channel base level. The inclined bedding surfaces of the bedforms are observed in both flow-parallel and flow-perpendicular sections, indicating their three-dimensional nature. Channel widening, channel meandering or both may result in an increase of channel bar bedform sand lateral accretion in portions of the channel as shown in Figure 7.53.

148

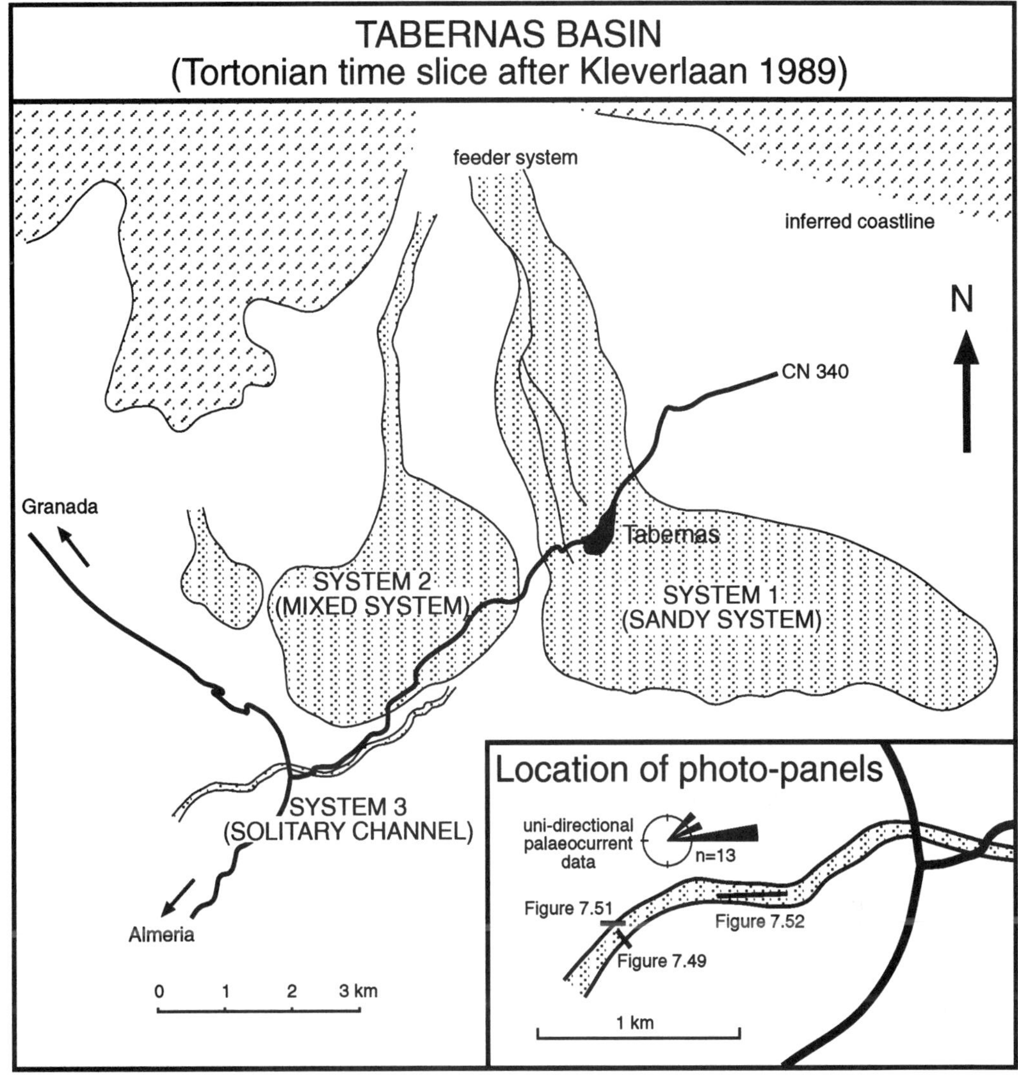

Figure 7.46. Tortonian time-slice showing the three turbidite systems. Redrawn after Kleverlaan (1987).

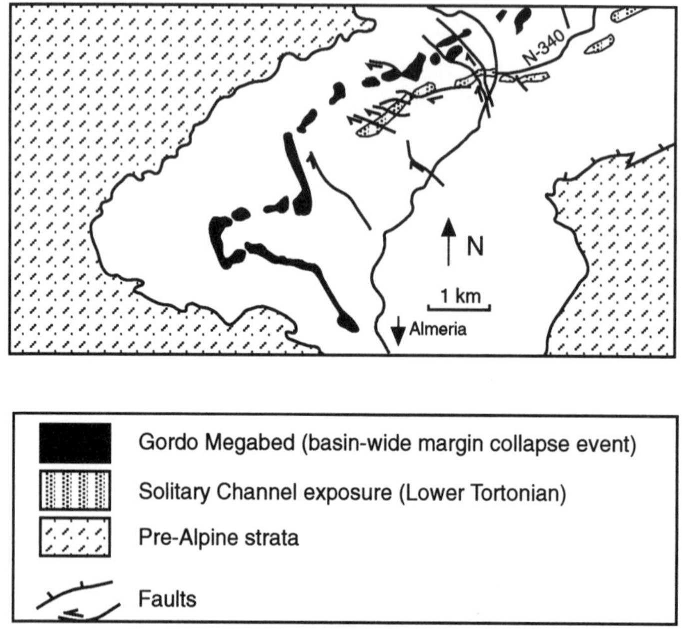

Figure 7.47. Map of the Solitary Channel outcrops and faults, some of which show evidence of being active during deposition, and hence are likely to have controlled channel course. Redrawn after Cronin (1994).

Figure 7.48. Cross-channel view of the Solitary Channel near its southern margin showing numerous faults. (See text for discussion).

Figure 7.49. (a) Photograph of the southern margin of the Solitary Channel, and **(b)** line-drawing interpretation showing the geometry of beds in the channel and towards the channel margin.

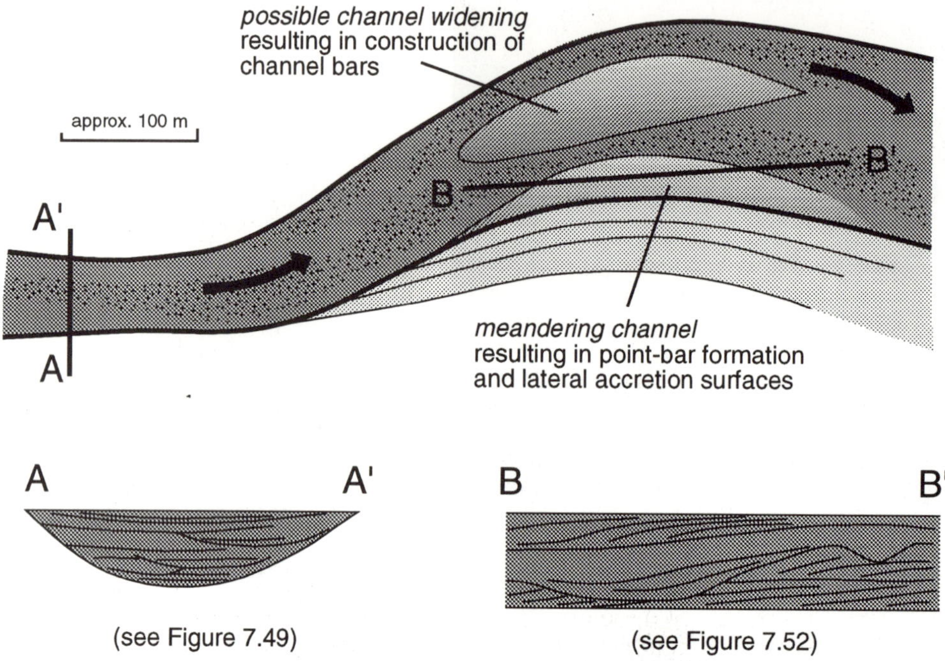

Figure 7.53. Schematic depositional model for the Solitary Channel outcrops west of the Granada-Almeria road. the geometry and outcrop relationships of the channel elements can be explained by either channel meandering (resulting in lateral accretion) or channel widening (resulting in intra-channel bar formation).

Namurian Shale Grit Formation slope-channel fills, Alport Castles, northern England

Age	Namurian
Fan position	Slope
Width	80 - 100 m (ch. element)
Depth	50 m (ch. complex).7 m (ch. element)
Length	-
Aspect Ratio	11 - 14 (ch. element)
Fill Type*	D
Max. Sinuosity	-

(see Figure 7.37)*

The Central Pennine Basin of north England developed in the Lower Carboniferous, and underwent its main clastic fill during the Namurian (Walker 1966a). The Namurian basin-fill sediments have been described and interpreted by Walker (1966a, 1966b), and Collinson (1968, 1970). The formations within the Namurian *Reticuloceras* zone demonstrate a general regressive sequence through the Edale shales (basin plain mudstones), Mam Tor Sandstones (distal turbidites), Shale Grit formation (proximal fan turbidites), Grindslow Shales (slope environment with sand-filled channels), and Kinderscout Grit (deltatop and fluviatile channels) (Figure 7.54). Walker (1966a) describes submarine channels from both the Shale Grit Formation and the Grindslow Shales. One of the most spectacular channel outcrops is exposed east of the River Alport in crags known as Alport Castles (Figure 7.55). The exposed sections are in cliffs up to 50 m high and almost parallel to palaeoflow direction. The facies are predominantly medium- and thick-bedded sandstones, showing a lack of grading (Facies C of Walker 1966a). Beds are commonly amalgamated and show numerous erosive bases. Using the architectural element scheme, the outcrop can be divided into 10 distinct second-order architectural elements, containing packets of beds, separated by thin silts or connected by down-cutting erosive surfaces (Figure 7.56). The entire cliff section may therefore represent the fill of a large-scale third-order channel or canyon, cut into fine-grained slope sediments, and contains second-order sand-filled channel and sheet architectural elements. One such channel element (#2) shows well developed terraced margins cutting into thin-bedded turbidites and siltstones (Figure 7.57). The channel-fill is made up of relatively few beds (5 maximum), and is slightly asymmetrical (see Figure 7.56). The top of the sequence ends with the abrupt change from sandy turbidites to laminated siltstones and mudstones.

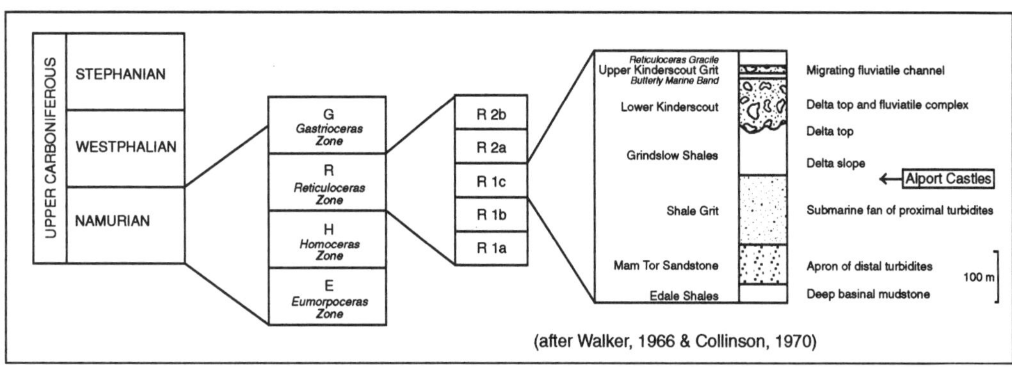

Figure 7.54. Upper Carboniferous stratigraphy. Redrawn after Walker (1966) and Collinson (1970).

155

Figure **7.55.** Main cliff section at Alport Castles (note figure for scale). The section is orientated almost parallel to palaeocurrent direction. Packets of beds can easily be traced across the outcrop (see Figure 7.56).

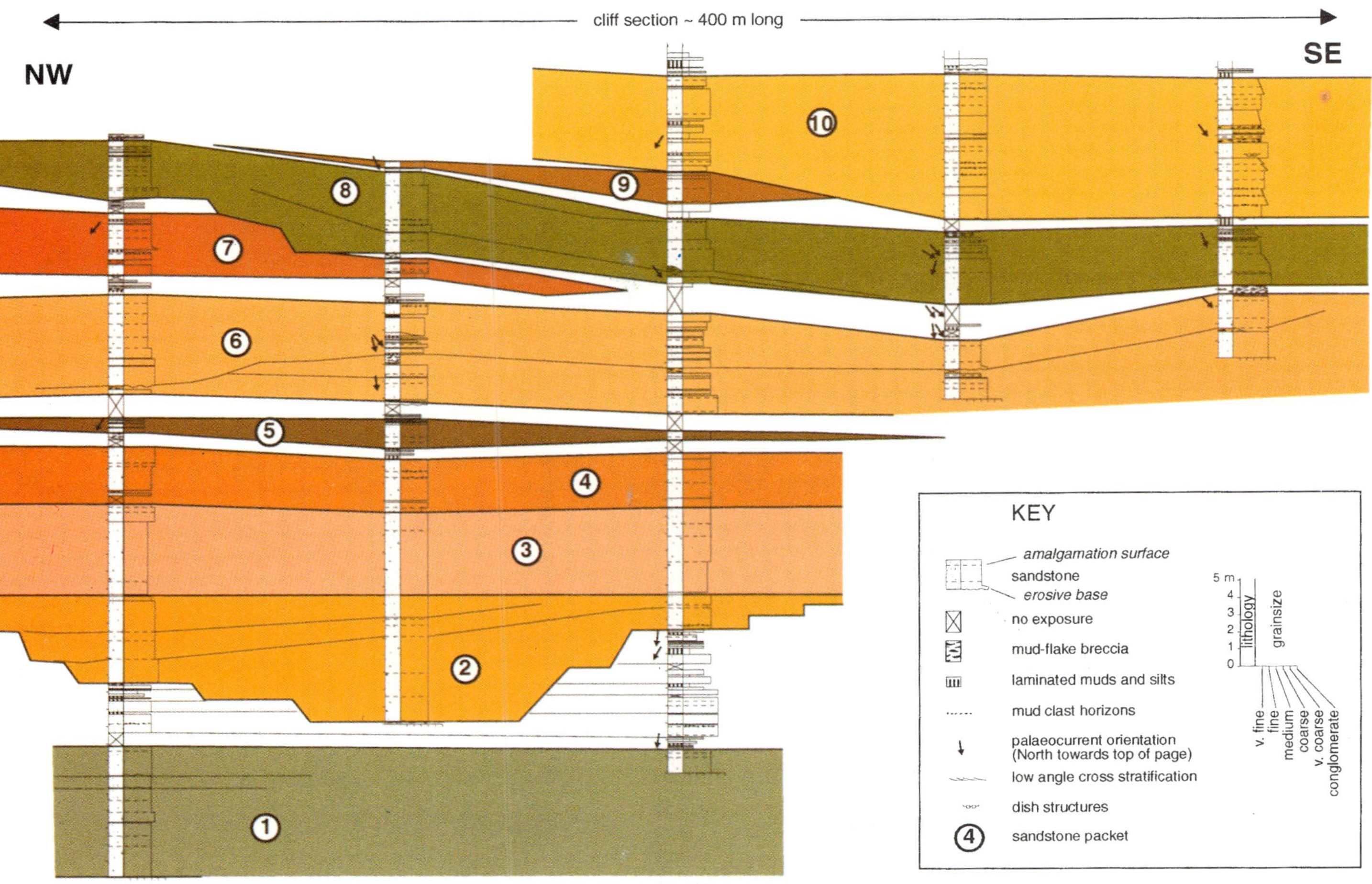

Figure 7.56. Correlation diagram of the Alport Castles outcrop shown in Figure 7.55 (Drinkwater & Clark in prep.). The correlation diagram shows individual sand packets, separated by mudstone intervals or prominant bedding planes. These sandstone packets are interpreted as channelised bodies. Note that true horizontal distance is hard to establish due to the very oblique orientation of the cliff with respect to general palaeocurrent direction.

Figure 7.57. View of the Alport Castles outcrop approximately perpendicular to palaeocurrent. Both margins of the channel and their stepped morphology can be seen (channel base arrowed). Note figure for scale above the arrow on the right.

Sealers Bay Channel, Oligocene Balleny Group, SW New Zealand

Age	Oligocene
Fan position	Middle fan
Width	1500 m (min. ch. complex).50 - 100 m (ch. element)
Depth	40 m (ch. complex). 10 m (ch. elements)
Length	-
Aspect Ratio	37.5 (min. ch. complex). 5 - 10 (ch. element)
Fill Type*	C
Max. Sinuosity	-

(see Figure 7.37)*

Chalky Island is situated at the entrance to the Chalky Inlet glacial fiord, SW Fiordland (Figure 7.58). The mountainous Fiordland National Park area largely consists of granulite and amphibolite grade regional metamorphic and acid-intermediate plutonic rocks. Chalky Island, however, is made up of Oligocene Balleny Group sediments, which also outcrop in small areas on the nearby mainland coast. The Balleny Group represents flysch basin formation along the western side of the New Zealand continental block (northern end of the Solander Basin) and shows a general transgressive sequence (Carter & Lindqvist 1975). The Puysegur Formation contains conglomerates breccias and arkosic sandstones and has been interpreted as "shelf" sediments (Carter & Lindqvist 1975). The Nuggets Formation consist largely of marine breccias, interpreted as deposits emplaced by inertia-flow in a submarine canyon. The overlying Sealers and Munida Formations consisting of sandstones and conglomerates have been interpreted as submarine fan sediments, and are overlain by the Chalky Island Formation chalk marls and thin-bedded turbidites which represent distal fan and associated abyssal seafloor deposits. The whole sequence has been compared to the facies and depositional environments in the Rio Balsas canyon system and associated littoral facies off western Mexico (Carter & Lindqvist 1975, see paper by Reimnitz & Gutierrez-Estrada 1970).

The Sealers Bay Channel is a spectacular erosive feature exposed in the Sealers Formation on Chalky Island (Figure 7.59). The channel-fill contains coarse-grained poorly-bedded sandstones, breccia and conglomerate lenses, and large (up to 150 cm) intraformational mud clasts. The channel is cut into similar facies with relatively steep

margins. Carter and Lindqvist (1975) describe the channel as being one of five amalgamated channel features in the continuous cliff exposure, which may therefore represent a channel complex at least 40 m thick (Figure 7.60). Credence to this interpretation is given by the overall thinning- and finning-upward sequence seen in the outcrop. Flow processes for the deposition of these facies are discussed by

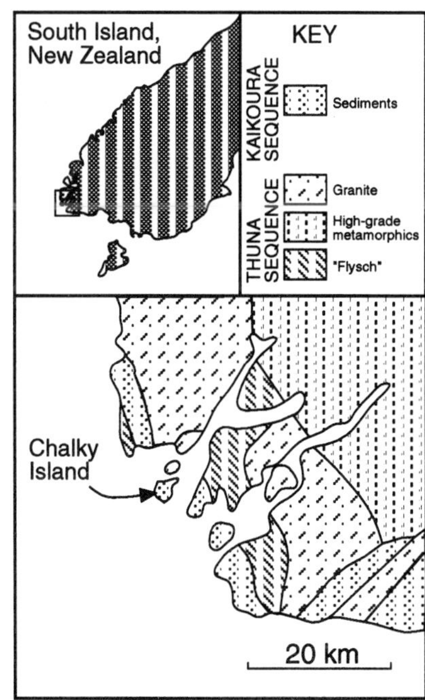

Figure 7.58. Simplified geological map of the Southwest Fiordland province, showing the location of Chalky Island. Redrawn after Carter & Lindqvist (1975).

159

Figure 7.59. Steep-sided scour element within the Sealers Bay Channel. The cliff face is approximately 12 m high. The channel-fill element is approximately 10 m thick and contains numerous large angular mud clasts which suggests a very local source and rapid single phase of scour infill.

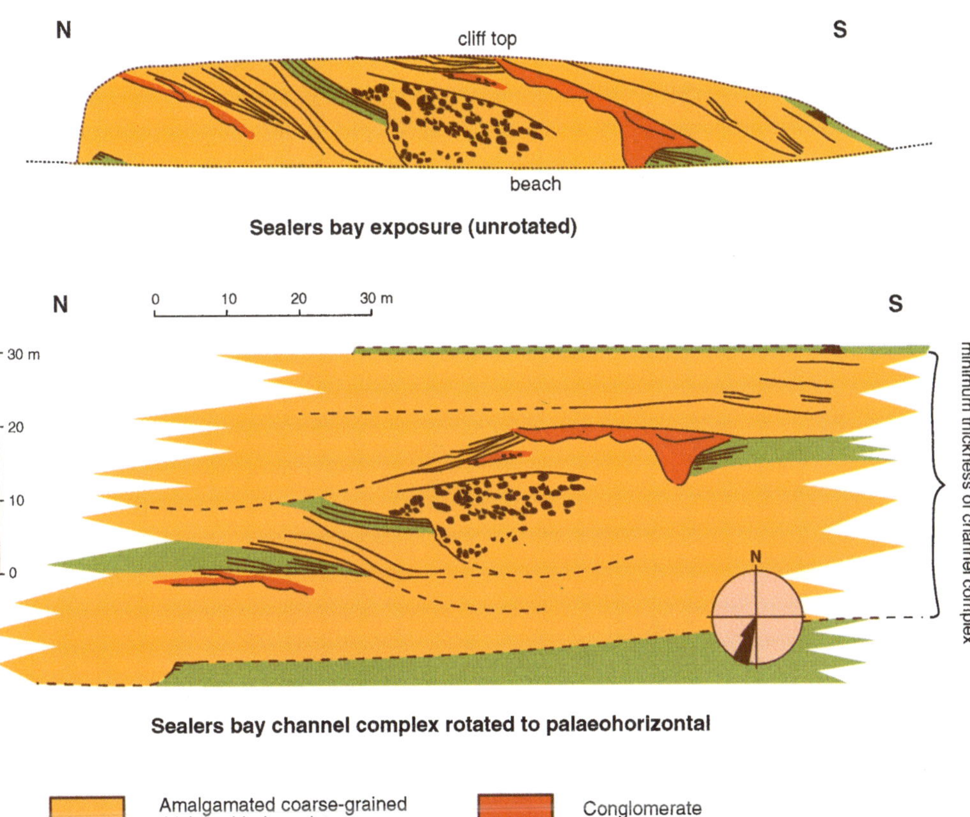

Figure 7.60. Sketch section of the Sealers Bay Channel outcrop, Chalky Island, SW New Zealand. The lower diagram is a rotated interpretation of the exposure, showing a channel complex of 40 m minimum thickness. Sedimentary data and panels produced by Pickering & Clark.

Carter and Lindqvist (1975). They interpret many of the channel-fill facies as being the deposits of 'flow-inertia' processes, including sand-flows and slide-creep, giving rise to diffuse contacts between sandstone and conglomerate deposits (Figure 7.61).

The Sealers Bay Channel is interpreted as a middle fan channel, due to the lack of levee deposits that may be expected in an upper fan environment. If such flow-inertia processes were predominant in the cut and fill of an erosive channel, then levees would not necessarily be formed. Palaeoflow direction was from the north towards the south and south-south-east. The channel is inferred to be relatively close to the sediment source, due to the high-energy nature of the flows and the mineralogical immaturity of the deposits (Carter & Lindqvist 1975).

161

Figure 7.61. Steep-sided conglomerate-filled scour elements cutting into thick-bedded sandstones. Note the diffuse contacts between conglomerate and sandstone.

Late Precambrian Kongsfjord Formation Hamningberg Channel, Finnmark, N. Norway

Age	Late Precambrian
Fan position	Middle fan
Width	500 m
Depth	20 m
Length	-
Aspect Ratio	25
Fill Type*	D
Max. Sinuosity	-

(* see Figure 7.37)

The 3.2 km-thick late Precambrian Kongsfjord Formation submarine fan system shows very well-developed channelised deposits interpreted as middle fan by Pickering (1982a). This section is based on various papers by Pickering (1979a, b, 1981, 1982a, b, 1985). The Kongsfjord Formation has been interpreted in all published work by Pickering as a large passive-margin fan analogous to the Mississippi Fan. In unpublished work by Pickering and Drinkwater, the formation is now reinterpreted as an essentially aggradational (not progradational) system that probably developed in a subsiding slope basin or several basins of much smaller size than originally proposed, and in which the stratigraphy resulted from the metronomic switching on-and-off of coarse clastics. Rather than a large-radius fan, the Kongsfjord Formation is perceived as analogous to the very thick (up to 1000s m), laterally-restricted (10+ km), growth-fault and diapir controlled turbidite basins associated with the continental slope from the present Mississippi Delta to Mississippi Fan or perhaps some of the northern North Sea Jurassic deep-marine systems. Each phase of sand input resulted in the basin-wide deposition of sheet-like sandbody up to 50-60 m thick, but typically 10-30 m thick, commonly associated with channelisation. These sand bodies are still referred to as "middle fan deposits" for ease of cross-referencing to the existing literature and because of their scale, but it should be borne in mind that no larger channel-levee-overbank complexes exist at outcrop in the Kongsfjord Formation.

The Hamningberg Channel is a good example of a channel margin showing lateral facies changes from channel-fill sandstones to levee deposits. At the eastern margin of the channel, there is a 6m-thick packet of thin and very thin-bedded, fine-grained, Bouma T_{cd} turbidites with some T_b divisions present (Figure 7.62). Locally, part of the channel packet shows evidence of sediment sliding.

Deposits in the eastern margin of the channel thicken and coarsen laterally until, in the western exposure, there is a 17 m-thick packet of very thick- to medium-bedded, granule- to very coarse-grained, sandstones and small-pebble conglomerates in beds that are invariably amalgamated (Figure 7.63).

Although individual beds cannot be correlated, a gradual thickening, coarsening and amalgamation of the channel deposits occur from east to west. For example, immediately west of the coast road, there are thin-bedded, coarse-grained, T_{bcd} turbidites that to the west become medium-grained, granule-grade/very coarse-grained T_{ab} turbidites. Towards the channel-axis, an erosive step in the channel margin can be seen with a packet of beds above representing a second-order slide element on the steep-sided terraced channel margin (Figure 7.64). Farther west, the turbidite beds amalgamate into small-pebble/granule/very coarse-grained sandstones and conglomerates, dominated by Facies Class A (Figure 7.65).

In the 17 m-thick part of the channel, palaeocurrents are oriented towards the south-east, whereas those from the 6 m-thick part are towards the east and east-north-east. (see Figure 7.63) This divergence is interpreted as showing deposition from turbidity currents that swung around from a down-channel direction to ascend the north-eastern margin of the channel and begin to construct levees. The 6 m-thick packet (see

Figure 7.62. Packet of thin and very thin-bedded, fine-grained sandy-silty turbidites interpreted as part of the inner channel-levee. Locally, part of the packet shows evidence of sediment sliding. Kongsfjord Formation, Finnmark.

Figure 7.62) may well represent the inner channel-levee deposits. The slide deposit, in the lower part, passes into undisturbed beds about 1 m thick over a lateral distance of about 30 m. The slide may have occurred as a result of deposition upon too steep a slope associated with the channel margin/inner channel-facing levee environment.

In the thickest development of this channel, the base is sharply erosive and the channel was excavated into mudstones, probably of a fan-abandonment phase. The most axial channel-fill deposits exposed show an overall thinning-and-fining-upward sequence. In the eastern exposure, however, erosion is not evident but, instead, there is a step-like thickening-and-coarsening-upward sequence. These two sequences are visualised as complementary; during the channel-infill phase, aggradation within the channel caused a shallowing with a corresponding increased propensity for onlap onto the inner channel margin and also overspill by flow-stripping. Thus, as the channel-axis thinning-and-fining-upward sequence was being generated, so too was the channel-margin/levee coarsening-and-thickening-upward sequence formed. Thus, both sequences represent complementary responses to one process - that of channel filling.

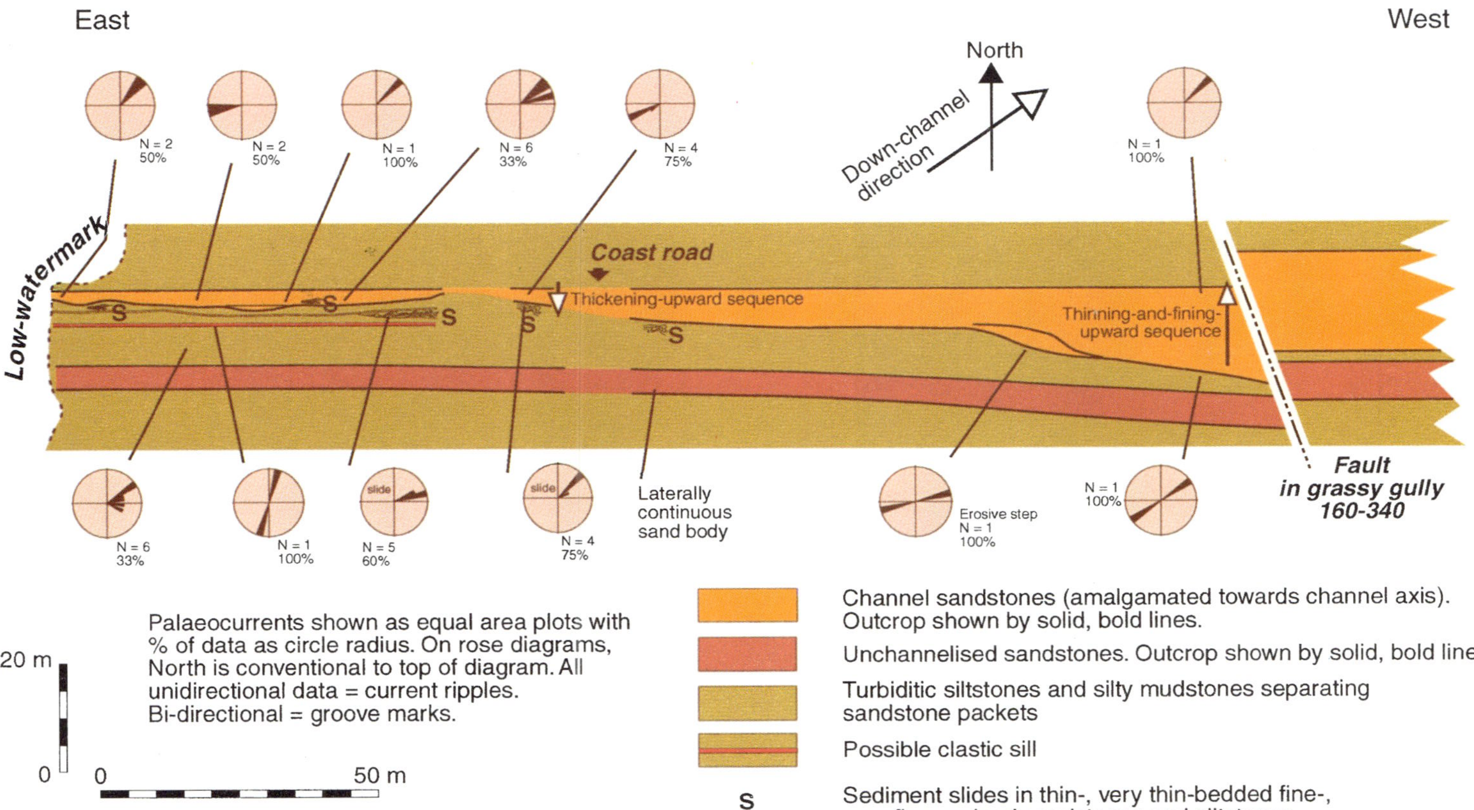

Figure 7.63. Measured section through the Hamningberg channel-margin, showing the architectural relationship between channel axis and channel margin deposits.

Figure 7.64. Erosive step down at the base of the Hamningberg Channel margin, with associated overlying slide deposit. Late Precambrian Kongsfjord Formation, Finnmark, northern Norway.

Figure 7.65. Axial-fill of the Hamningberg Channel. Most beds are amalgamated coarse- to medium-grained sandstone turbidites. Late Precambrian Kongsfjord Formation, Finnmark, northern Norway.

Brushy Canyon Formation, Middle Permian, Guadalupe Mountains, Texas

Age	Middle Permian
Fan position	Upper Fan
Width	1000 m (Salt-flat Channel), 400m (100-foot Channel)
Depth	45 m (Salt-flat Channel), 30 m (100-foot Channel)
Length	-
Aspect Ratio	22 (Salt-flat Channel), 13 (100-foot Channel)
Fill Type*	Fill type: D
Max. Sinuosity	-

(see Figure 7.37)*

Several submarine channels are exposed on the western flanks of the Guadalupe and Delaware mountains, westernmost Texas. The exposures form part of a shelf-basin transect through the Middle Permian Brushy Canyon Formation. The slope to basin-floor transition outcrops over about 24 km. Figure 7.66 shows the scale and type of exposures of the Brushy Canyon Formation and an unnamed sand-filled channel incised into siltstone and thin-bedded turbidites (slope facies).

A subsurface continuation of the Brushy Canyon Formation constitutes the producing fields in the Delaware Basin some 100 km to the north-east. Most of the channels in the Brushy Canyon Formation can be classified as either Type C or Type D, both of which show at least two different sedimentary processes; initial flows responsible for generating an erosional channel, and later flows responsible for the infilling of the channel (Zelt & Rossen 1995). Figure 7.67 shows an example of an erosive and stepped channel margin cutting into largely interchannel facies. The fill of the channel consists of thick-bedded amalgamated sandtones, resulting in an extremely high sand:shale ratio.

The Brushy Canyon Formation has been interpreted as an overall lowstand depositional system (Silver & Todd 1969, Sarg & Lehmann 1986). Water depths in the basin have been estimated as 50 to over 300 m based on the observed relief of submarine erosional unconformities (Zelt & Rossen 1995). The uppermost Brushy Canyon Formation may have been deposited in depths of 50 - 150 m due to the infilling of the basin (Zelt & Rossen 1995). The two submarine channel outcrops selected here are known as the "Salt-flat" channel and the "100-foot" Channel.

The "Salt-flat" Channel

The "Salt-flat" Channel is located at the top of the Brushy Canyon Formation. Three-dimensional exposures allow a detailed examination of the channel axis and north-eastern channel margin (Figure 7.68). The base of the channel is marked by a slide-scar feature, which is in-filled by sandy debrites and high-density gravity-flow deposits containing floating clasts, traction carpet lags etc. The internal channel architecture shows a complex stacking of second-order sand-rich channel, sheet and lateral-accretion elements (see Figure 7.68b). At least four offset stacked coarse-clastic elements constitute the fill of the channel (Sonnenfeld & Gardner 1996, Gardner & Sonnenfeld 1996).

The "100-foot" Channel

The "100-foot" channel is located at the top of the Brushy Canyon Formation. The channel is filled with vertically stacked amalgamated thick-bedded turbidites which become unamalgamated towards the channel margins (Zelt & Rossen 1995). The channel has an asymmetrical cross-section which can be viewed almost perpendicular to mean palaeoflow direction in the outcrop shown in Figure 7.69. The diverging palaeocurrents recorded in this channel suggest that the "100-foot" channel was slightly sinuous (Zelt & Rossen 1995).

167

Figure 7.66. View of an unnamed sand-filled channel from the upper Brushy Canyon Formation, Guadalupe Mountains. The channel is viewed towards the north-east (i.e. the section is orientated obliquly to palaeocurrent direction). The channel cuts into siltstones and thin-bedded turbidites (slope facies). The channel margin can clearly be seen in the centre of the photograph. The plate also shows the scale and type of exposure in the area.

Figure 7.67. Erosional stepped channel margin in the Brushy Canyon Formation. The channel is filled with amalgamated thick-bedded sandstones.

(a)

(b)

Figure 7.68 (a). View of the "Salt-flat" Channel beneath El Capitan (carbonate facies). Palaeocurrent direction is approximately towards the viewer and the channel axis has been eroded left of the prominent mesa. **(b)** Close up of the "Salt-flat" Channel showing the geometry of intrachannel elements. Palaeocurrent direction is approximately towards the viewer, and the channel axis is towards the left. From the channel margin (right) towards the axis the beds can be seen to thicken and amalgamate above a clearly defined erosive channel base.

Figure 7.69. View of the "100-foot" channel (arrow) upper Brushy Canyon Formation. The channel is cut into thin-bedded turbidites and laminated siltstones, and filled with thick-bedded turbidites (*cf.* Zelt & Rossen 1995)..

Gryphon Field channels, Upper Paleocene, North Sea

Age	Upper Paleocene
Fan position	Slope/ Upper fan
Width	500 m - 1000 m
Depth	7 m - 120 m
Length	-
Aspect Ratio	5 - 70
Fill Type*	C & D
Max. Sinuosity	1.23

(see Figure 7.37)*

Several channel sand bodies have been found in the Quad 9 area of the northern North Sea in the Balder Formation (Upper Paleocene), and the Frigg Formation (Lower Eocene). The channels largely contain facies B1.1, B2.2 with some Facies C2 and F2.1 (facies classification of Pickering *et al.* 1986a, 1989) (Timbrell 1993). The sands are interpreted as being formed from high-density channelised turbidity currents. Three time slices have been constructed, constrained by biostratigraphy, allowing the mapping of channel sand bodies.

The two most striking attributes of the channels that have been mapped, are large down-channel thickness variations and the lateral stacking of sand bodies across a sediment fairway. Down-channel thickness-changes range from 120 m to 30 m and 50 m to 7 m over distances of 1 km. These thickness variations and the slightly sinuous channel geometry gives the reservoir sands the appearance of a "meandering string of pearls" (Timbrell 1993). Timbrell interprets this sand body geometry to flow-stripping processes that may occur in sinuous channels.

The lateral stacking of sand bodies appears very similar to channel stacking of the Ainsa II channel. The channels appear to be confined within sediment fairways, but show diachronous lateral stacking patterns (Figure 7.70). These channels are interpreted as forming a constructional (or aggradational) system on a prograding slope (Timbrell 1993). Timbrell suggests that the modern day Valencia margin off north-east Spain is a reasonable analogy to the Balder Formation depositional system.

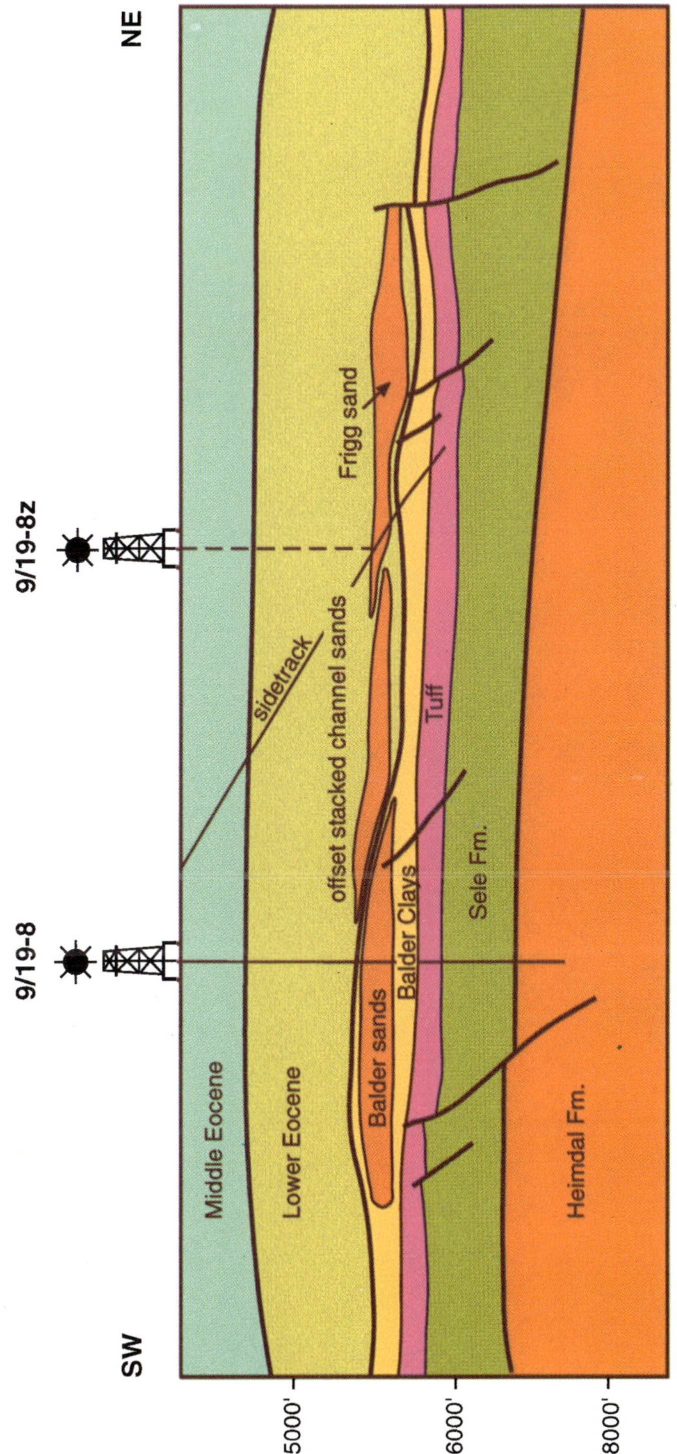

Figure 7.71. Balder Channels in the Gryphon Field, northern North Sea showing offset stacked channel architecture (Timbrell 1993).

Chapter 8

Quantitative analysis of ancient channel complexes

Dataset

Ancient channel width and depth measurements are commonly taken as the measurements of the coarse-grained part of a channel-complex. Although these measurements are likely to approximate to the channel dimensions in most cases, it must be remembered that the channel-fill may also contain fine-grained "abandonment facies" which will result in an under-estimation of true channel depth. Channel width can also be estimated from channel half-width measurements, where exposure does not allow both channel margins to be seen. Channel width and depth parameters therefore can easily be obtained from outcrops with sufficient exposure, allowing aspect ratios and cross-sectional areas to be calculated.

Ancient channel planform geometry in outcrops can rarely be mapped with the degree of accuracy necessary to calculate sinuosity and meander geometry. The use of 3-D seismic in hydrocarbon exploration, however, can provide detailed maps of the planform geometry of channel features. Accurate measurements of channel slope are virtually impossible to ascertain in ancient channel systems due to post-depositional compaction and tectonic tilting. It is therefore not feasible to examine the relationship between slope and sinuosity of ancient channels.

Channel dimensions

Channel width and depth measurements have been taken from the submarine channels case studies of Chapter 7 and from ancient channels described in the literature. Figure 8.1 shows a plot of channel width and depth. The spread of data on the log-log plot is similar to that of modern channel width and depth, although the data is biased towards channels with smaller dimensions. A comparison of width/depth plots between modern and ancient channels is discussed in Chapter 9. Ancient channel width:depth ratio varies from 3.5 to over 100. Larger values of width:depth ratios, however, may be difficult to recognise in outcrop.

Aspect ratios of intra-channel elements

Detailed measurements of the width and depths of intra-channel elements and individual beds have been taken from the Ainsa II Channel Complex (see Chapter 7). Such measurements are only possible from outcrops with relatively good exposure, where detailed correlation of individual beds and packets of beds has been achieved (see Figure 7.34). In addition to using the measurements from "complete" elements (i.e., those which are observed in their entirety within the limits of the outcrop and correlation certainty), a method is proposed

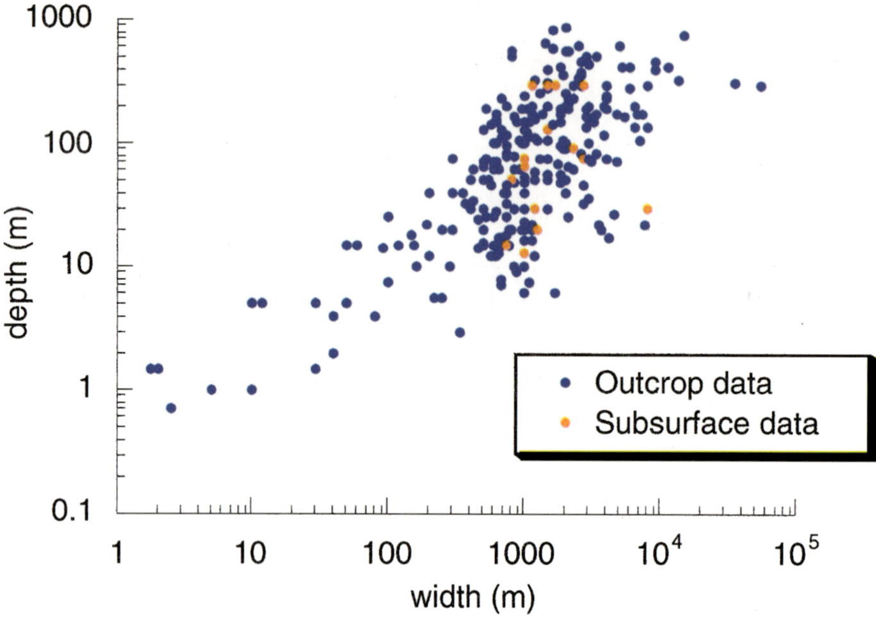

Figure 8.1. Width:depth ratios of submarine channels, with subsurface channel data plotted for comparison.

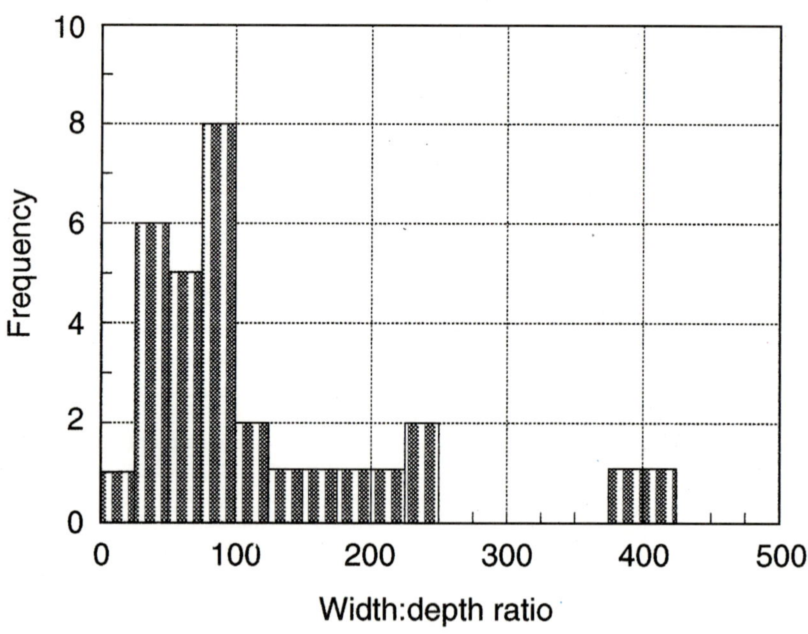

Figure 8.2. Histogram of aspect ratios (width:depth ratios) of the intra-channel elements of the Ainsa II Channel Complex (refer to Chapter 7 and Figure 7.34).

176

here to estimate widths of "partial" elements to substantiate the dataset. An approximation of *element width* can be gained by measuring the rate of element thinning in both directions to determine the estimated *pinch-out lengths*. The *estimated element width* is simply the addition of the two *pinch-out lengths*. The estimated element width, therefore, can only be calculated when thinning is observed in both directions, i.e. an assumed maximum depth has been observed in the outcrop. Of course the calculation of element width assumes a constant rate of thinning and does not consider the fact that many channel elements are later eroded by subsequent intra-channel erosion surfaces. Therefore the *estimated element widths* are likely to be largely overestimated.

Figure 8.2 shows a histogram of the aspect ratios of intra-channel elements within the five offset stacked channels of the Ainsa II Channel Complex. Both "complete" elements and "estimated" elements are included, and all widths have been corrected to widths perpendicular to palaeocurrent, from apparent widths measured at outcrop. The elements considered also range from individual beds to second-order intra-channel elements. The frequency distribution shows that the intra-channel elements largely have greater aspect ratios than the channel bodies in which they are deposited in. There also appears to be a polymodal distribution to the aspect ratios of these elements (which may be expected since all types of intrachannel elements have been used in this analysis) although more data points would be required to substantiate this trend. Most of the intra-channel element aspect ratios lie between 25 and 100, whereas the channel bodies (numbered 1-5 on Figure 7.34) all have aspect ratios less than 40. This may be an expected finding, but the quantification of such differences between the aspect ratios of channels and elements within channels is of great importance to the application of stochastic modelling techniques to subsurface submarine channel reservoirs.

179

Part IV

Synthesis

&

Submarine Channel Models

Chapter 9

Synthesis

Architectural elements from selected examples of modern and ancient channels are identified to aid interpretations of channel-fill type, the interconnectivity of channel bodies resulting from channel stacking, causal sedimentary processes, and architectural relationships between channels and levees. Architectural elements are described using the classification scheme presented by us in Pickering *et al.* (1995b).

Channel-fill type

The channel types of Normark (1970) (erosional, depositional and erosional-depositional or mixed), and their related fill deposits, can be identified in both modern and ancient settings (see Figure 1.5). This submarine channel classification scheme is useful for modelling channel sedimentation with respect to intra-basin, tectonic and sea-level controls, and autocyclic channel switching.

The Eocene Ainsa I channel complex, southern-central Pyrenees, is a good example of a mixed erosional-depositional channel (Mutti & Normark 1991). The good exposure of the channel-fill and channel margin make it ideal to demonstrate how the identification of architectural elements can be used to classify channel type. The channel fill sequence commences with a residual conglomeratic, medium- and thick-bedded sandstone element, overlain by deposits resulting from one or more debris flows. These elements are associated with the initial erosional phase of the channel, and its early

depositional fill, where the majority of the sand grade sediment was transported through the channel and deposited farther downslope in extra-channel sites. The overlying deposits, constituting the bulk of the channel fill, relate to the depositional phase of channel development, blocking the channel by back-filling processes, interpreted as due to base-level rise (Mutti & Normark 1991). This predominantly medium-bedded sandstone channel element, although largely depositional in character, is locally eroded into the underlying muddy debrite element. Figure 7.29 is an architectural element interpretation showing how this section of the channel consists of smaller architectural elements related to the change in depositional style from erosional to depositional. The coarse-grained sandstone beds of a second-order channel element, interpreted as a channel thalweg, show clear onlap onto the underlying deposits. Channel thalweg overbank elements pond to the south of the thalweg margin, in a similar fashion to larger scale interchannel sand elements. The ponded elements are locally eroded by smaller channel elements, containing lateral accretion elements, showing a shift in the channel thalweg axis. The overlying sheet sand element is composed of many small and broad stacked channels with low-angle lateral accretion surfaces above the thalweg associated deposits, and this element represents the later-stage depositional type of the channel infill sequence.

Deep-towed high resolution side-scan instruments can reveal 2-D plan views of architectural elements of the order of scale that could be identified from outcrop. The EEZ-SCAN 84 GLORIA survey of the Monterey fan channel off the western coast of the United States shows that the sinuous upper fan channel is fed by submarine canyons located at the shelf edge near Monterey Bay, California, and can be mapped southwards for 300 km (EEZ-SCAN 84 Scientific Staff 1986). A high resolution side-scan sonar survey of the Monterey Channel, undertaken in 1990 using the TOBI deep-tow system, has delineated the sedimentary features of the present day channel-fill (Masson *et al.* 1995) (Figure 5.2). Figure 5.3 shows an image interpretation of a portion of the TOBI survey, and the main architectural features of the channel. The second-order architectural elements within the channel include, high-backscatter lobe elements, a high-backscatter bedform field, irregular elements of mixed-backscatter facies flanking the channel margins (interpreted as levee elements), and second-order scour elements, including the waterfall-like cut downs, channel terraces and pitted-low-backscatter overbank scour elements. The architectural elements identified may be used to characterise the type of sedimentation occurring within parts of the channel in the most recent past. The presence of many scour elements and residual high-backscatter bedforms within the channel indicates that this section of the channel exhibits largely erosive-type channel sedimentation (*cf.* Normark 1970).

A similar suite of architectural elements occurs in the Ashgill-Llandovery Milliners Arm Formation, north-central Newfoundland. From detailed facies logging and mapping, Watson (1981) describes deep-marine, sand-dominated, submarine channel deposits showing axial (thalweg) reworked bedforms (lag deposits, very irregular coarse-grained sandy beds, and abundant decimetre-scale cross-stratification), channel-margin terraces and associated deposits, finer grained levee-overbank deposits, and sandy crevasse channel sediments (Figure 9.1). We suspect that these features are typical of relatively straight, low-sinuosity, channel systems, and the architectural elements of the channel fill are comparable in geometry and scale to those observed within the present-day Monterey channel described previously (see Figures 5.2, 5.3).

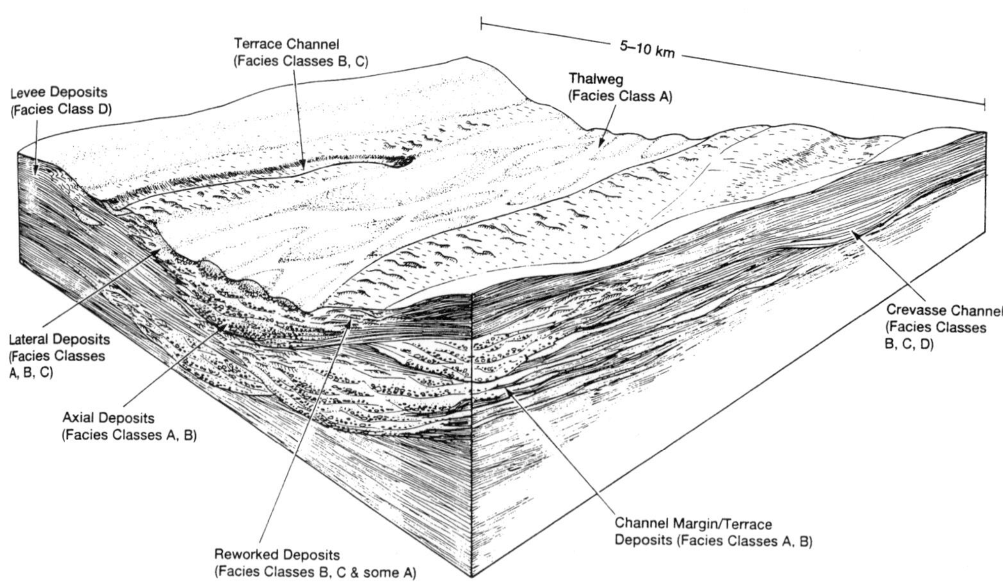

Figure 9.1. Interpretative block diagram of the upper fan channel deposits of the lower Silurian (Llandovery) Milliners Arm Formation (Watson 1981). Facies classes are those of Pickering *et al.* 1986a, 1989).

182

Channel stacking

The stacking architecture of channel sand bodies observed in outcrop or from seismic section, demonstrates channel growth patterns, forming from the interaction between lateral and vertical amalgamation processes operating during the growth of the channel system (Figure 9.2). Some of these processes are discussed below. The effect of channel growth pattern has a strong control on the interconnectivity of individual channel-fill elements and the width:depth ratio of reservoir bodies comprising stacked channel-fill sand bodies (Figure 9.3).

The Late Miocene San Clemente submarine deposits of the Capistrano Formation provide a well documented example of the sequential lateral shifting of channelised deposits. The channel-fill facies consist of "classical" turbidites, structureless sandstones, pebbly sandstones and mudstone drapes. The nested submarine channels, described and numbered by Walker (1975c), can be defined using the architectural element scheme of Pickering *et al.* (1995b) as third-order sand, pebbly sand and mud-filled channel elements. The sedimentary fill of each third-order channel element, however, may be characterised by smaller scale observations, for example, Channel 1 contains a second-order sheet sand architectural element, and the second-order sandy inclined geometry of the lowermost exposed fill of Channel 4 may be interpreted as a lateral accretion element. Third order channel elements showing sequential offset stacking relationships, such as the nested channels of the Capistrano Formation (Walker 1975c), indicate progressive lateral shifting in the channel axis, due to either channel migration or avulsion.

The well studied Ainsa II channel of the Eocene Hecho Supergroup, in the south-central Pyrenees, is another example of a channel showing third-order offset stacked channel bodies (Clark & Pickering 1996) (see Figures 7.33, 7.34). The progressive southwards stacking of these channel elements has been interpreted to have resulted from channel avulsion caused by tectonic tilting of the basin floor (Mutti *et al.* 1989, Clark & Pickering 1996). In the Paleogene of the northern North Sea, the Lower Eocene Balder Sands and Frigg Sand, above the Balder Clays, show an eastward/north-eastward lateral offset stacking of submarine channels (see Figure 7.70), which compares well in scale with the Eocene Ainsa II channel growth pattern from the Spanish Pyrenees. In a similar way in which the channel stacking of the Ainsa II channels resulted from tectonic control, the lateral offset in the Balder Sands channels also appears to have been controlled by faulting.

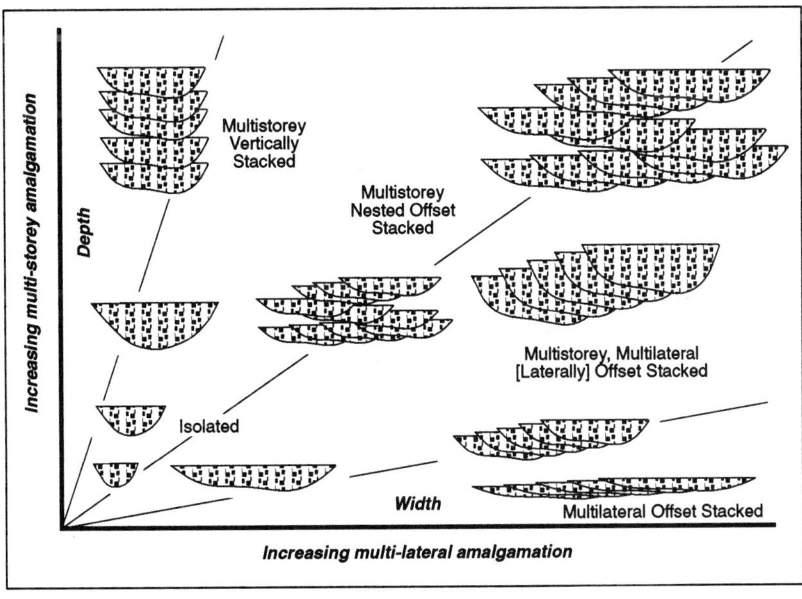

Figure 9.2. Schematic graph showing how the width:depth ratio of stacked channel complexes results from the interplay between lateral and vertical amalgamation of individual channel bodies. From Pickering *et al.* (1995b).

183

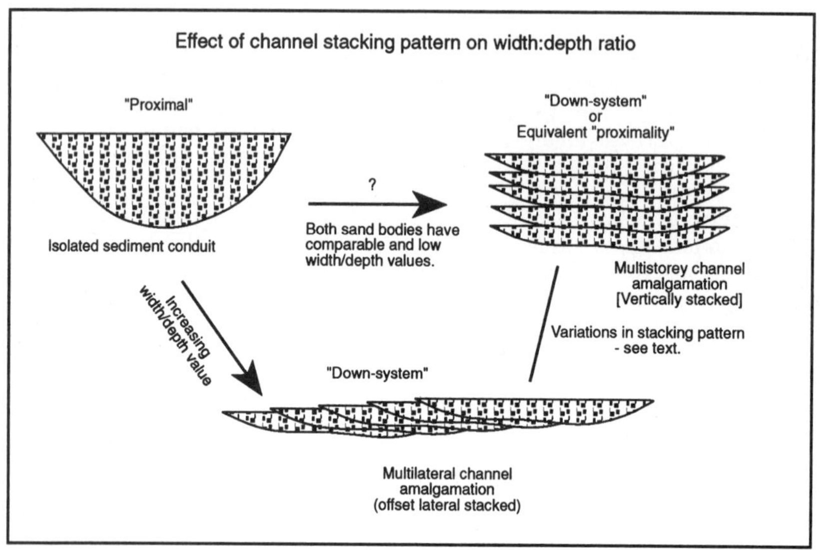

Figure 9.3. Diagram to show how the width:depth ratio of a reservoir body comprised of stacked san-filled channel elements depends on the growth pattern of submarine channel systems. From Clark & Pickering (1996).

The East Kongsfjorden turbidite channel of the late Precambrian Kongsfjord Formation, northern Norway contains second-order offset stacked channel elements (Figure 9.4). The stacking of these channel elements is interpreted to have resulted from the lateral migration of the channel (Pickering 1982a). Purely autocyclic processes that result in a shift of channel axis can be modelled to demonstrate the stacking architecture and channel body connectivity. Reservoir models exist for fluvial channels demonstrating that the interconnectivity and distribution of channel bodies within the floodplain deposits are related to:

- Laterally variable aggradation.
- Compaction of fine sediment.
- Tectonic movement at floodplain margins.
- Channel avulsion (Bridge & Leeder 1979).

Vertical stacked channel architecture can result from the control of channel axis location by inherited basin floor topography, submarine canyons, salt, mud or igneous diapir activity, or synsedimentary faulting (Figure 9.5). Detailed correlation of the Llandovery Caban Conglomerate Formation, south-central Wales, constrained by good biostatigraphic control using graptolite subdivisions (Smith *et al.* 1991), demonstrate vertical stacking architecture of the channel

conglomerate elements, controlled by synsedimentary faulting (Figure 9.6).

The course of the Miocene "Solitary Channel" in the Tabernas Basin, SE Spain, was strongly controlled by a major syndepositional fault/fault-system, causing part of the channel to flow perpendicular to the palaeo-slope direction (Kleverlaan 1989). Field evidence shows that the fault was a normal growth fault during channel-fill sedimentation, with a seafloor step measured in metres. The channel course has been accurately mapped, and can be traced for at least 8 km showing a low channel sinuosity (Kleverlaan 1989). Recent work suggest that the "Solitary Channel" may be traced back further towards its feeder system over a total distance of 11 km (Cronin 1995), but we are yet to be convinced of this correlation (unpublished data). The sinuosity is a result of changes in channel course which are most likely to be structurally controlled (Kleverlaan 1989, Cronin 1995). The channel is at least 200 m wide and 40 m deep (Cronin 1995) which gives it an uncharacteristically low aspect ratio compared with other channels. The strong tectonic control makes it difficult to establish if exposed faulted margins of the channel represent true channel margins i.e. whether the fault was the actual channel margin, or if channel margin deposits have been subsequently displaced from axial-fill deposits by later movements on the .

184

Figure 9.4. East Kongsfjorden channel margin deposits, Late Precambrian Kongsfjord Formation system Finnmark, northern Norway. Cliff section in foreground is about 15 m high. Note the progressive cut-and-fill (arrowed) from right to left, i.e., offset-stacked elements, indicating lateral migration to the north (left-hand side) (Pickering 1982a).

Controls on vertically stacked channels and other sediment conduits

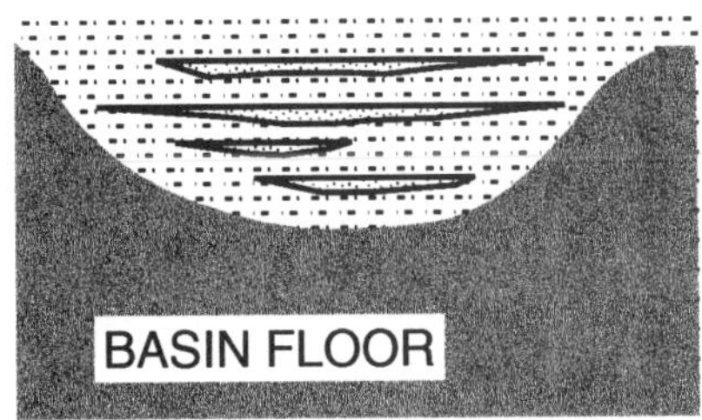

Inherited
Basinfloor Topography

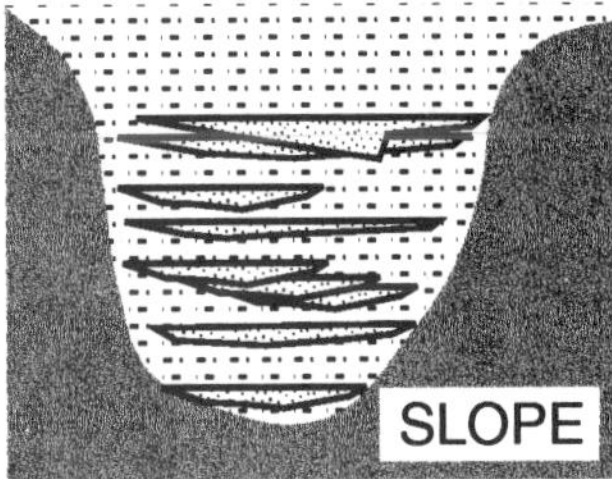

Intra-Canyon

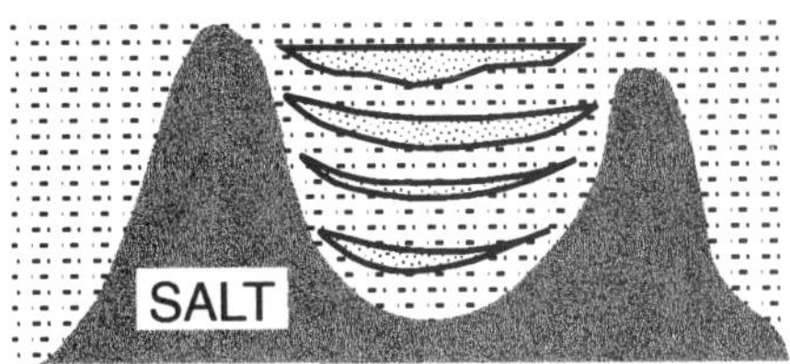

Diapir (salt, mud, igneous)

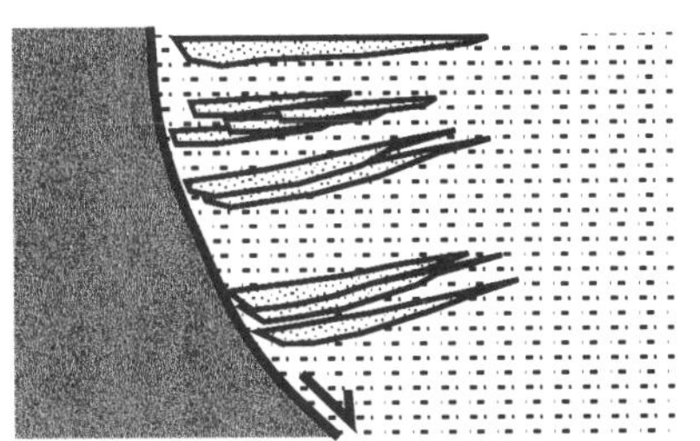

Growth Fault

Figure 9.5. Controls on vertically stacked channels and other sediment conduits. From Pickering *et al.* (1995b).

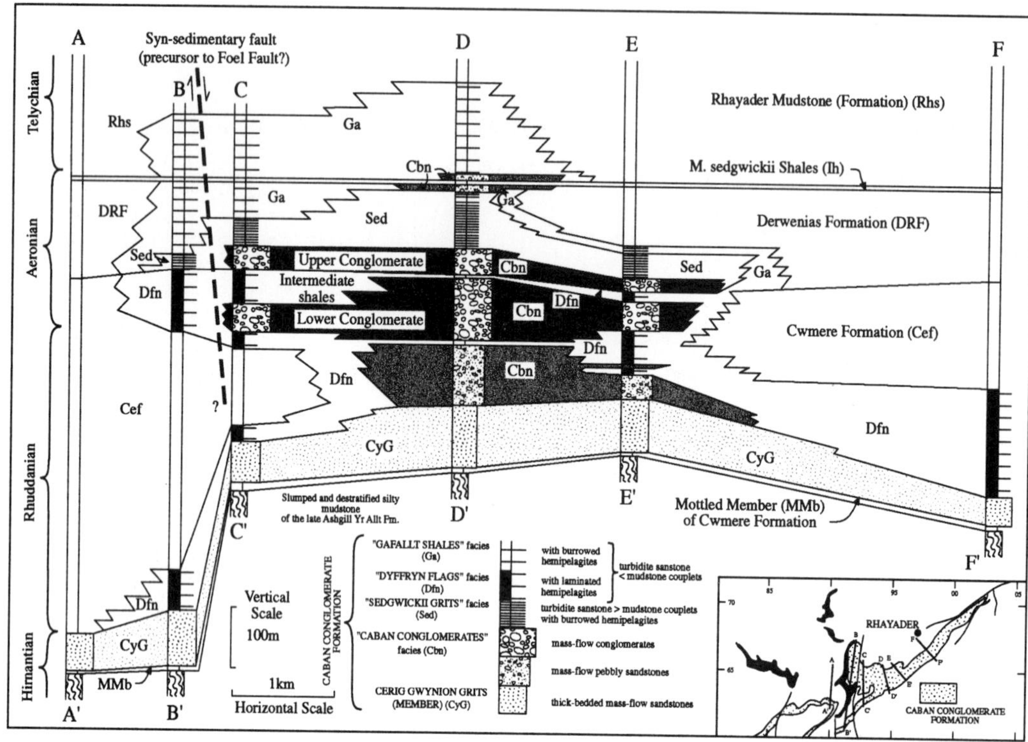

Figure 9.6. Architecture of the Silurian (Llandovery) Caban Coch Conglomerate Formation, Welsh Borderland, showing the fault-controlled vertical stacking pattern of the channels. Redrawn after Smith *et al.* (1991).

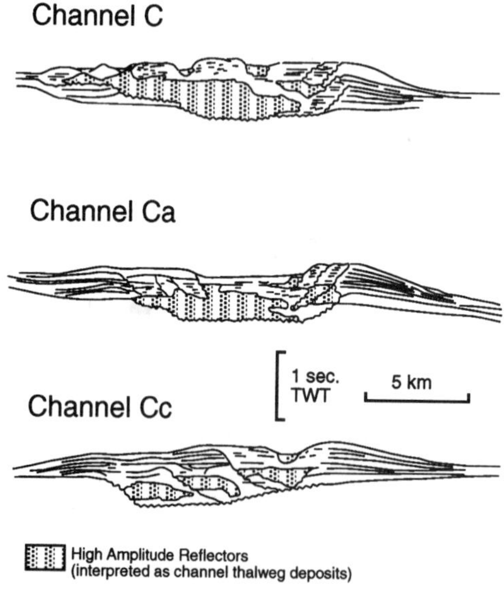

Figure 9.7. Lateral offset stacking architecture of three different channel-levee complexes, Indus inner fan channels. TWTs estimated as: Channel C = 400 m thick, with levees up to 350 m thick; Channel Ca = 400 m thick, with levees up to 200 m thick; Channel Cc = 550 m thick, with levees 275 m. The channels begin as erosional channels and evolve to essentially aggradational depositional channel-levee-overbank complexes. Redrawn after McHargue (1991).

fault/fault system. Whatever the precise geometry of the "Solitary Channel", it is a good example of a low-sinuosity sediment conduit controlled by growth faulting along kilometres of its reach.

McHargue (1991) documents a variety of stacking patterns for channels on the Indus Fan in the area equivalent to canyon complex 2 of Kenyon *et al.* (1995a). Many of these channels are offset stacked, and some appear to have a main infill which is interpreted from the seismic character as sandy, and all appear to have an erosional base. Figure 9.7 shows three large-scale erosional-depositional channel-levee complexes in which smaller scale channel axis (thalweg channel) elements are stacked with different degrees of lateral continuity and vertical connectivity. Within all three channels McHargue (1991) describes the decrease in lateral

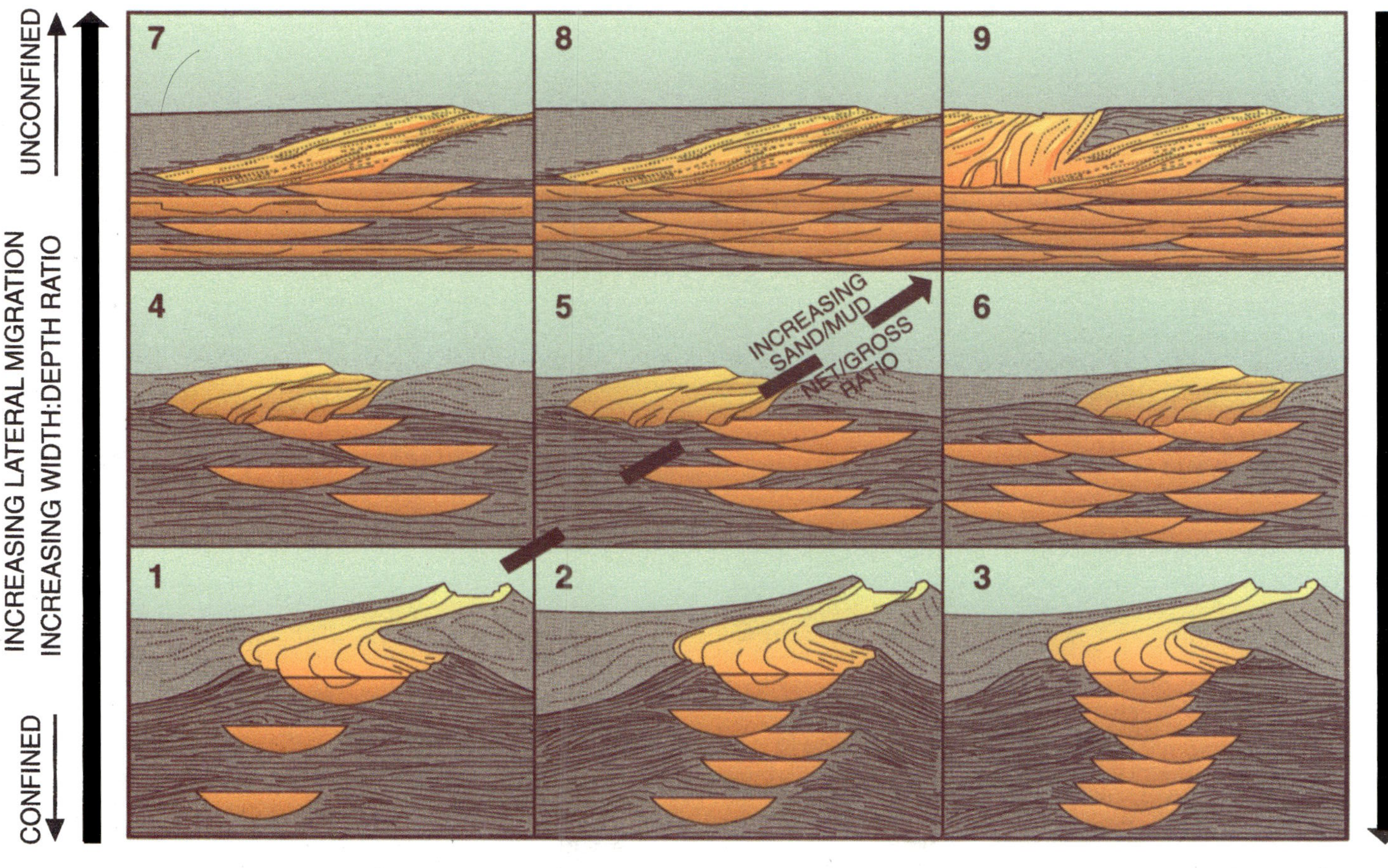

Figure 9.8. Schematic model to show a variety of channel stacking patterns (numbered 1 - 9), and how the stacking of channel bodies is likely to be related to sinuosity, rate of sediment accumulation, sand/mud ratio of sediment supplied (net/gross ratio) and the degree of channel confinement.

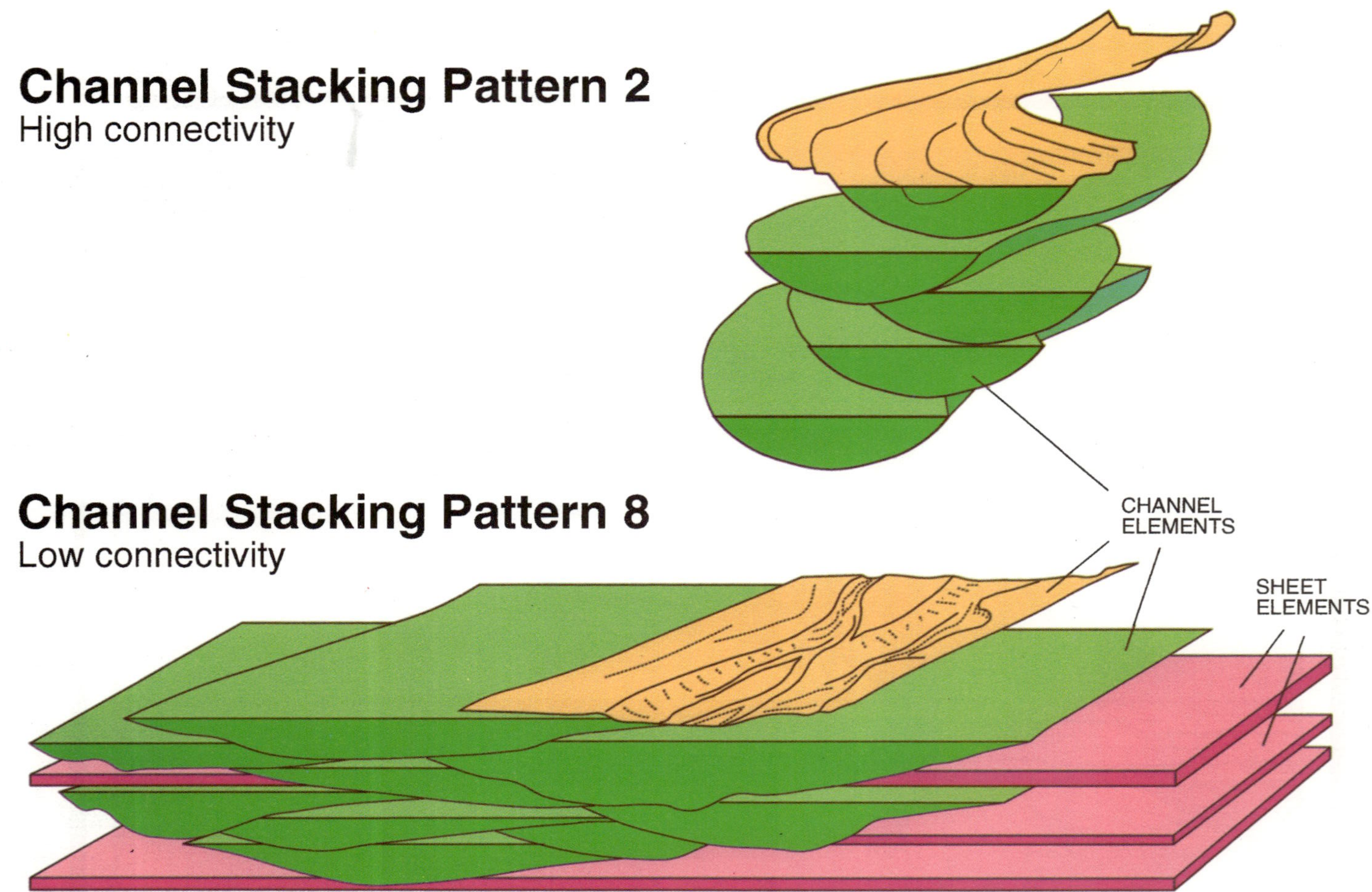

Figure 9.9. Schematic examples of channel stacking patterns 2 and 8 from Figure 9.8, showing sand-bodies "lifted-out" from surrounding sediment. The diagram contrasts the difference between the aspect ratios and connectivity of the sand-bodies.

continuity and thickness of the (sandy) channel axis elements during channel development, commensurate with a decrease in the thalweg meander belt. In all cases, these changes are related to an overall decrease in channel lateral migration and an increase in channel aggradation.

Figure 9.8 shows a schematic diagram of channel stacking patterns numbered 1 - 9. The diagram shows how autocyclic channel stacking is likely to be strongly related to channel type. The development of levees in a low sand:mud ratio system will promote a strong degree of vertical aggradation, whereas in sandy systems, channels can be expected to have a strong degree of lateral migration due the absence of mud-rich levees. Channel stacking pattern is also likely to be strongly controlled by the rate of sediment accumulation. Figure 9.9 shows two examples of channel stacking patterns, and how the channel stacking pattern controls the interconnectivity of sand-filled channel elements. Relatively unconfined channels systems will have a high degree of channel body interconnectivity (e.g. channel stacking pattern 8) unlike confined vertically aggraded systems (e.g. channel stacking pattern 2).

The extreme case for the stacking of channels with relatively little confinement results in the stacking of more sheet-like elements (as shown in part in the channel stacking pattern 8 on Figure 9.9). An example of such a system can be seen in the Montagne de Chalufy Channel of the Eocene Grès d'Annot Formation (Figure 9.10) (Hilton & Pickering *in prep.*). The channel elements of this channel are almost sheet-like, passively infilling and onlapping onto the margins of a large channel-erosion surface.

Figure 9.10. Large-scale channel erosion surface (50 m deep) showing sheet-like sandstone bodies onlapping the channel margins, Montagne de Chalufy Channel, Eocene Grès d'Annot Formation, south-east France Hilton & Pickering 1995, *in prep.*, photograph courtesy of H.D. Sinclair).

Sedimentary processes

Modern submarine channels commonly display sinuous or meandering patterns, controlled by the behaviour of through-channel flows. On large-scale modern channels, such as the those on the Indus, Amazon and Mississippi fans, features can be observed which appear similar to features formed from fluvial hydraulic processes, with evidence of frequent channel avulsion and channel migration (Figure 9.11). Many of the spectacular features associated with modern submarine channel architecture, for example ridge-and-swale topography, do not appear to be present in the ancient rock record - but, this may be a problem of scale, or simply lack of careful documentation. Although small-scale lateral accretion deposits occur in ancient channels, large-scale lateral accretion elements (similar in size to the scroll bars on the Mississippi Channel, Pickering *et al.* 1986a), are largely absent from ancient channel fill sequences (Mutti & Normark 1991). A reason for this disparity between modern and ancient channels may be due to the unrecognition of low-angle lateral accretion surfaces in outcrop, but may also be related to differences in the type of basins that are commonly preserved. Apart from accretionary prisms, most of the ancient rock record in deep-marine systems represent the vestiges of upper continental slope, intra-shelf, or aulacogen-related deposits where basins are up to orders of magnitude smaller than continental margins and ocean basins,

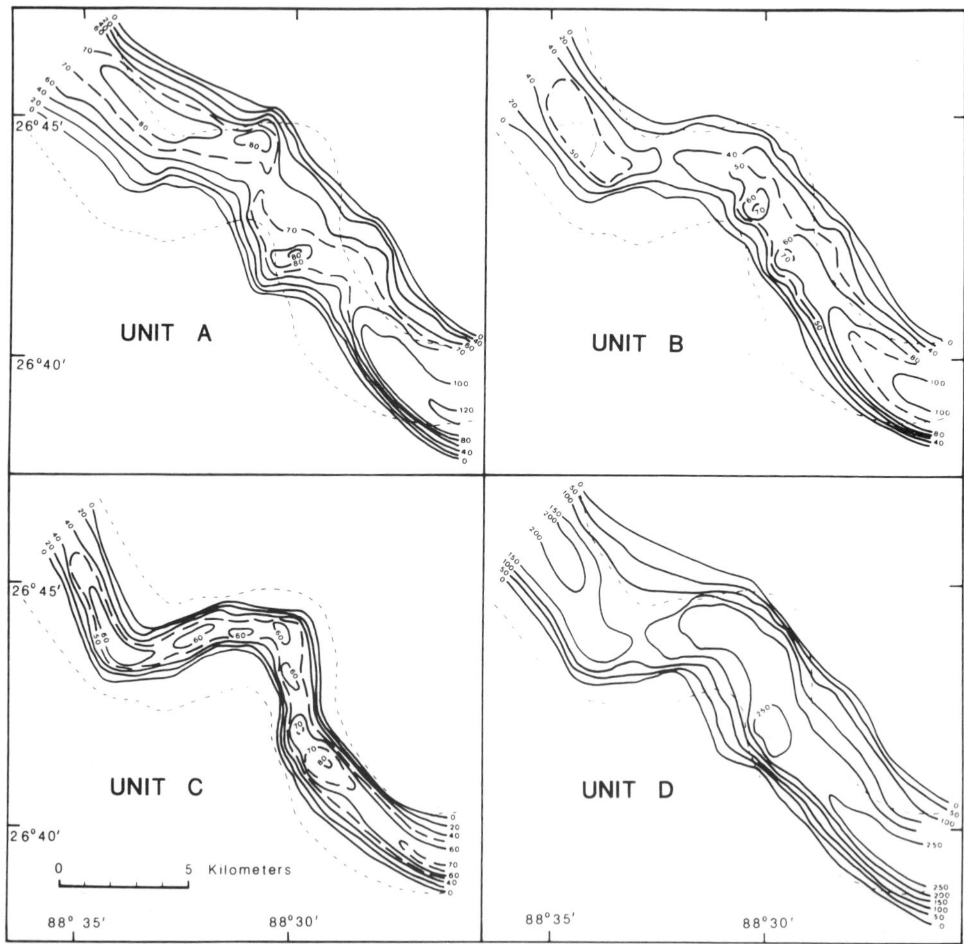

Figure 9.11. Isopach maps of high-amplitude reflectors showing how the Mississippi channel thalweg has migrated during its recent evolution. Channel thalweg units A-C are shown oldest to youngest with the levee crests of the present day channel by the dashed line. Contoured intervals are in milliseconds (Pickering *et al.* 1986b).

190

with few if any essentially flat basin floors, and where slope gradients are commonly high. A corollary of this is that erosional and erosional-depositional, low- to moderate-sinuosity, channel systems should prevail.

Dipmeter data has lead to the interpretation that the channel-fill of the subsurface Permian Indian Draw Field, New Mexico contains sandy offset-nested mound elements (Phillips 1987). The geometry and architectural elements suggest meandering processes were involved in channel deposition, although the scroll-bar features are a factor of approximately five times smaller than those seen in the Mississippi channel-levee complex. The nested mounds of the "A3" Sandstone interval are progressively stacked from the inner channel bend margin into the channel axis, in concordance with deposits formed from lateral accretion processes (Figure 9.12). The nested mounds of the "A1" Sandstone interval, however, are progressively stacked from the outer channel bend margin (Figure 9.13). We suggest that these nested mound elements have been formed from flow stripping processes. Flow deceleration resulting from a momentum change

associated with a change in turbidity-current flow direction at a bend may lead to flow stripping processes, thereby, causing a reduction in the turbulent energy of the current and the partial deposition from suspended load: channel bar elements consisting largely of the coarsest grained parts of the sediment load may result (Piper & Normark 1983). The architectural elements produced by these processes may make excellent intrachannel reservoir rocks, since as well as consisting of coarse-grained, high-porosity material, they will inherently contain reworked and "clean" sediment, due to the nature of this type of depositional process. Stratigraphic seals may result from such sandy elements being enveloped by mud-rich levee and intrachannel deposits, formed during phases of channel abandonment and, or, relative inactivity. The implications for petroleum exploration lie in the prediction that good reservoir bodies may be located at or near channel bends, although such reservoirs may consist of individual lenses of sandstone bodies separated by permeability barriers, and probably offset stacked in a down-channel pattern due in part to compensation stacking sensu Mutti and Sonnino (1981).

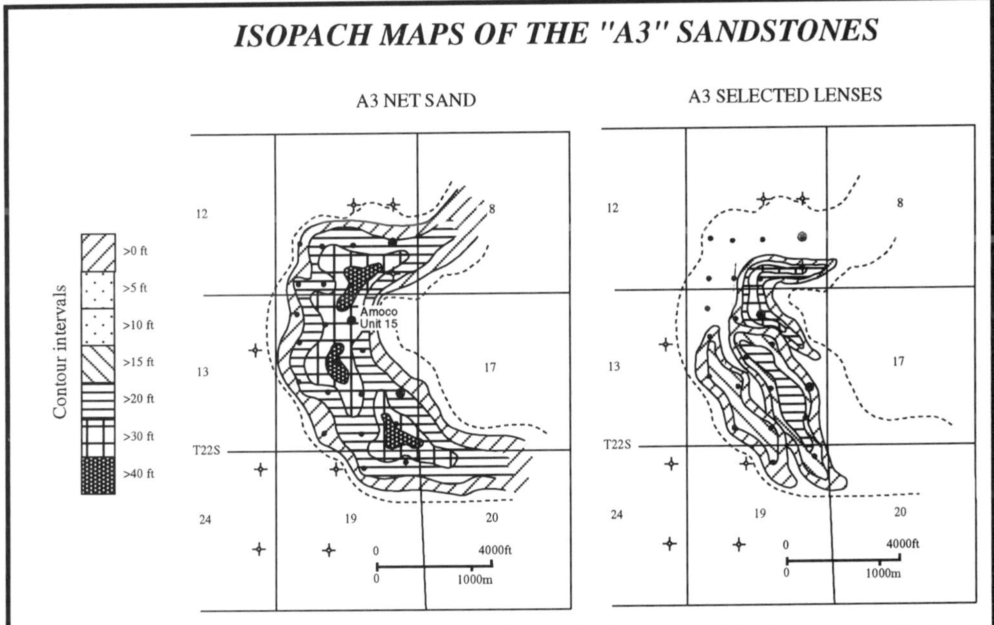

Figure 9.12. Isopach maps of the nested mounds of the "A3" Sandstone interval, progressively stacked from the inner channel bend margin into the channel axis, in concordance with deposits formed from lateral accretion processes, Permian Indian Draw Field, New Mexico. Redrawn from Philips (1987).

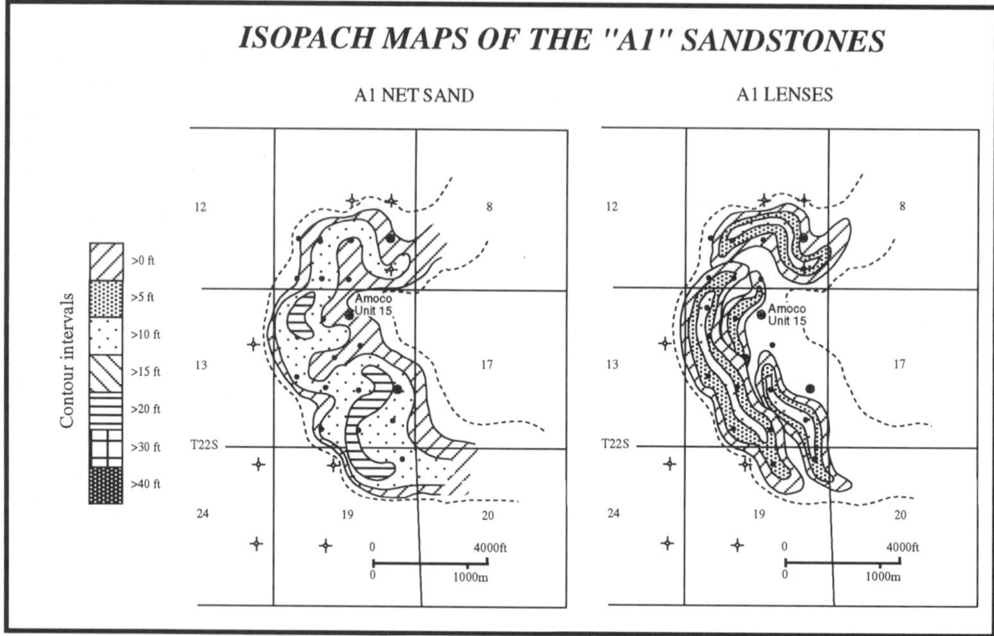

Figure 9.13. Isopach maps of nested mounds of the "A1" Sandstone interval, progressively stacked from the outer channel margin, Permian Draw Field, New Mexico. Redrawn from Philips (1987). We suggest here that these nested mound elements have been formed from flow stripping processes. See text for explanation.

Flow deceleration processes are manifest in the Umnak Channel, which feeds sediment into the Bering Sea basin from the Aleutian volcanic islands. The channel is an example of a low-input and low-sinuosity, but high-gradient, channel system (Kenyon 1992). GLORIA sonographs in the canyon/upper fan channel area show a 3-km-wide channel containing intrachannel second order high-backscatter mound elements, commonly located at or near the channel bends and confluences (Figure 9.14). These elements have been interpreted as relatively coarser-grained channel bar architectural elements, deposited from processes involving changes in flow velocity of turbidity currents flowing around channel bends (Kenyon & Millington 1995). The Umnak Channel demonstrates that these flow deceleration processes can occur in relatively low sinuosity channels.

Large-scale bedforms can be observed in the Miocene "Solitary Channel" of the Tabernas Basin, south-eastern Spain (see Figure 7.52). The bedforms mainly comprise thick-bedded (up to 4 m) coarse-grained sandstone and conglomeratic beds (Facies B1.1 and Facies A1.4) which pinch-out over distances of 5-150 m. These three-dimensional inclined-surface macroforms may be classified as having first- and second-order sigmoid architectural geometry (sensu Pickering *et al.*

1995). These bedforms could represent deposits of a lateral channel bar (e.g. lateral accretion element or offset stacked beds deposited as a result of flow stripping seen in this outcrop at an oblique angle. The scale of these features provides a good ancient example of modern sandy macroforms (such as sandwaves) that have been observed in modern channel systems using sidescan sonar (e.g. Malinverno *et al.* 1988, Hughes-Clark *et al.* 1990, Masson *et al.* 1995).

The formation of large-scale flutes, and groups of these features has been observed in modern channel settings (e.g. Hughes Clarke *et al.* 1990, unpublished data from western Mediterranean). Large-scale turbulent flows have been interpreted as being responsible for their formation. The scale of such features can be similar to channel dimensions of ancient successions. Figure 9.15 shows some large-scale flutes at the base of the feeder channel infill of the Sandy System, Tabernas Basin. It is possible that some documented ancient channels may be the infill of large-scale mega-flutes. Furthermore, subsurface maps of reservoirs bodies giving the appearance of a repetitively thickening and thinning channel-fill (commonly known as a string of pearls geometry) may alternatively be down slope groupings of infilled mega-flutes.

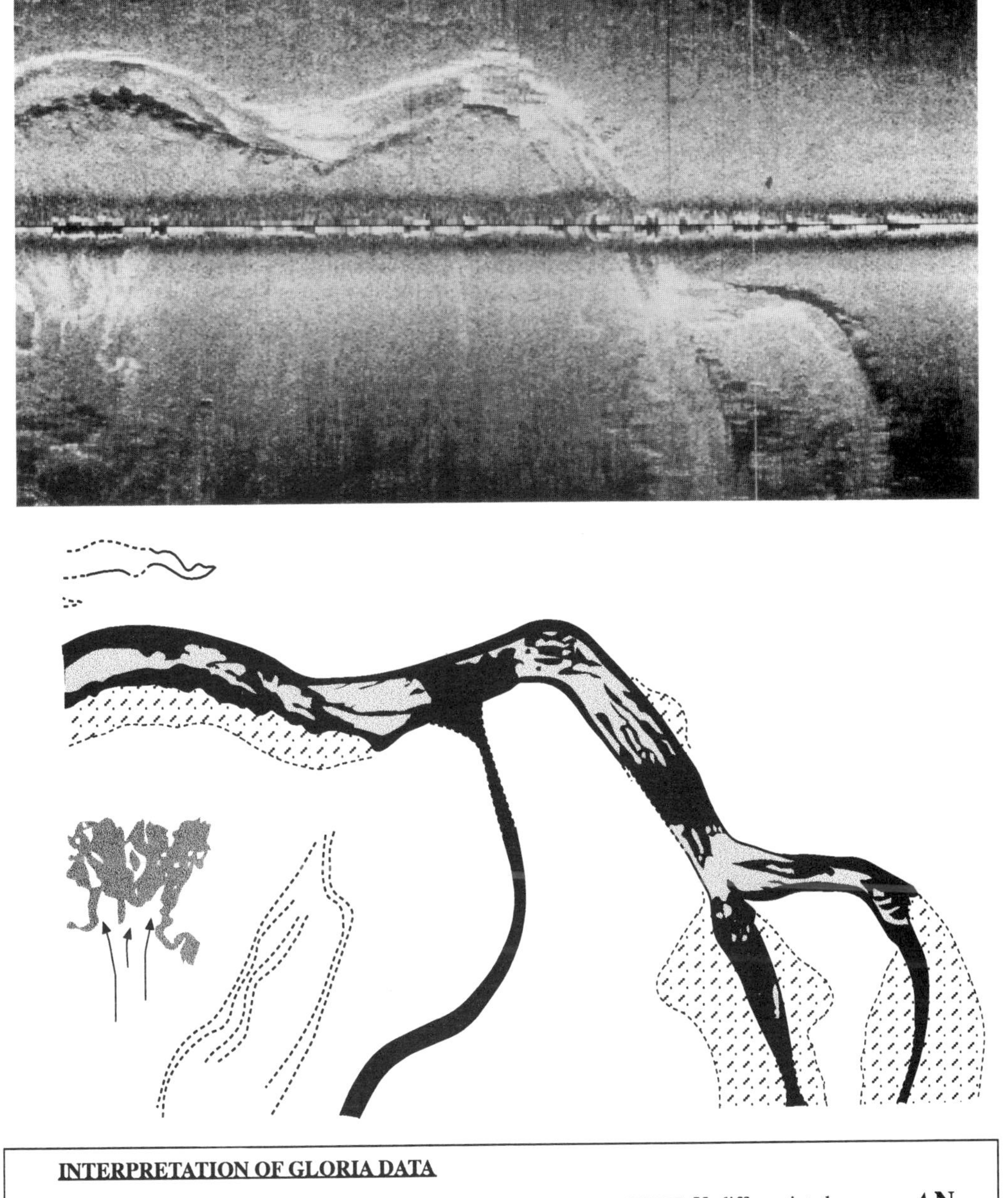

Figure 9.14. GLORIA sonograph and interpretation of a portion of the Umnak Channel, Bering Sea (Kenyon & Millington 1995). The sonographs shows the location of high-backscatter elements within the channel at locations including the channel confluence and channel bends. One interpretation is that these areas of high-backscatter are intra-channel bars.

Figure 9.15. Flow-parallel section of the the feeder channel to the Sandy System, Tabernas Basin. Flow is towards the left-hand side. At the base of the section, large-scale flutes can be seen, infilled with debris-flow deposits.

Relationships between channels and levees

Channel levees are formed from the overspill of channelised turbidity currents. On modern deep-marine channels, channel levees may be present, and can be easily identified on high resolution seismic profiles, showing positive relief (lens geometry), and commonly second order irregular and, or, bedform geometry on their surfaces, related to seafloor erosion and modification of levee crests by overbank currents. Seen in plan, channel levee architecture can generally be characterised by irregular or mound geometry (see the levee architectural elements on the Monterey channel, see Figures 5.2, 5.3). The effects of the Coriolis force result in higher right-hand-side of levees (looking down-channel in the Northern Hemisphere), with the opposite situation for levees in the Southern Hemisphere (Menard 1955, Komar 1969). Levees can be up to 50 km wide and 300 m above the surrounding abyssal plain (e.g. Amazon Fan, Damuth *et al.* 1988).

Ancient levee deposits possess a high potential for subsequent wet-sediment modification, e.g. differential compaction, growth faulting and inter-/intra-stratal slip, and are hard to distinguish from other basin-slope deposits. Because it is virtually impossible to preserve and recognise the positive relief morphology of levees in ancient channel complexes, ancient levees deposits are grouped into the more general facies-association term, "overbank" deposits, as described by Mutti and Normark (1987). Overbank facies comprise fine-grained and thin-bedded, current-laminated deposits, commonly together with hemipelagic/pelagic mudstones. Levee architectural elements may be identified within overbank deposits, but it is generally difficult to apply the architectural element scheme to such sequences.

The architecture and facies of modern and ancient levee deposits have been described by many authors (e.g. Normark *et al.* 1980, Pickering 1983, Damuth & Flood 1984, Pickering *et al.* 1986b, Kolla & Coumes 1987). Typically, levee deposits show large-scale bundling of beds, up to many metres thick, into laterally-pinching packets of beds, wedging out over hundreds of metres to a few km, e.g. as seen on the high-resolution seismic sections of the Mississippi channel levees (Figure 9.16). Levee sediments generally show rhythmic occurrences of laminated silty turbidite beds in the order of 1-5 cm thick, with some thicker, medium- to fine-grained sandy beds; bed geometry is commonly very irregular on a scale of metres. Inclined, contorted and folded laminae are common, resulting from wet-sediment sliding within gravitationally unstable accumulations of levee deposits. In some cases, turbiditic levee sediments may be interbedded with debris-flow deposits, and crevasse sands.

Levee sediments, including the associated hemipelagic/pelagic intervals, require much more research because they:

- Tend to record most, if not every, through-channel flow event.
- Provide, with high-resolution dating, an accurate record of the longevity of a channel as a conduit for the through-put of sediments.
- With continuous vertical recovery where pelagic sediments exist, may record changes in global climate and, or, other extrabasinal controls on sediment accumulation.

A commonly cited problem with many ancient channel-levee-overbank complexes (generally interpreted by analogy with modern middle-fan channels), is the inability to "walk-out" beds from within a channel into the levee or overbank deposits. Some rare exposures of ancient channels, however, do permit this, where sands appear to have onlapped high up the inner slopes, and possibly even have overspilled into interchannel areas. Figure 7.62 shows the architectural elements of the Hamningberg Channel, Kongsfjord Formation, north Norway (Pickering 1982a). The architecture of this third order channel contains both axial-fill elements in lateral continuity with channel margin and levee elements. The axial fill element (a second-order, small pebble-granule/very coarse-grained sandstone and conglomerate channel element) passes laterally into channel margin and levee architectural elements (e.g. a second-order slide element on the channel margin and a thickening- and coarsening-upward sequence), with no observed break in sedimentation. The medium-grained, thick-bedded, sheet element, containing palaeocurrents diverging away from the main channel current direction, may be a clastic sill/dyke architectural element.

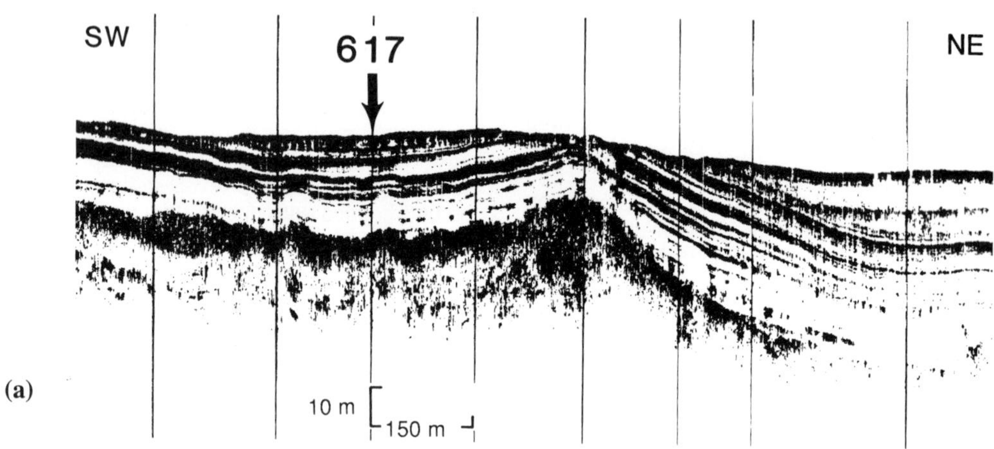

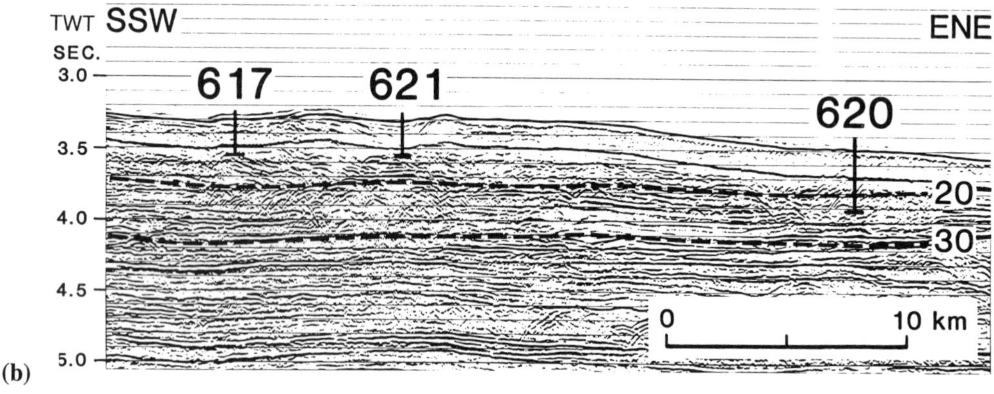

Figure 9.16 (a) 4.5 kHz high resolution EDO deep-towed seismic reflection profile over the channel margin and levee of the present day Mississippi Channel (Pickering *et al.* 1986b). **(b)** Multichannel seismic reflection profile across the Mississippi Channel. After Pickering *et al.* (1986b).

In contrast to the thickening-and-coarsening-upward sequence at the eastern channel margin, the axial-fill element in the western part of the channel fill shows a thinning-and-fining-upward sequence. A possible explanation is that the gradual infilling of the channel by vertical accretion resulted in the lateral progradation of the intrachannel deposits, away from the channel axis and onto the channel banks. This process generated a thickening-and-coarsening-upward sequence at the channel margin/levee. Thus the two sequences are seen as complementary, representing differing responses to one process, i.e. "sympathetic sedimentation" (Pickering 1982a).

Quantitative synthesis

One of the aims of this book is to establish quantitative relationships involving width and depth measurements to quantitatively define types of deep-marine sediment conduits. Width and depth measurements were measured from a variety of modern and ancient channels (including subsurface channel bodies), canyons, gullies and scours. Figure 9.17 shows the distribution of the data which approximates to a log-normal distribution (Figure 9.18). The graph shows a bias towards smaller features measured in ancient sequences compared to those from the modern seafloor. The general distribution of aspect ratios (width:depth ratios) however, is similar between modern and ancient settings. We conclude therefore, that there is no reason why aspect ratios from ancient channels should differ from those of modern channels.

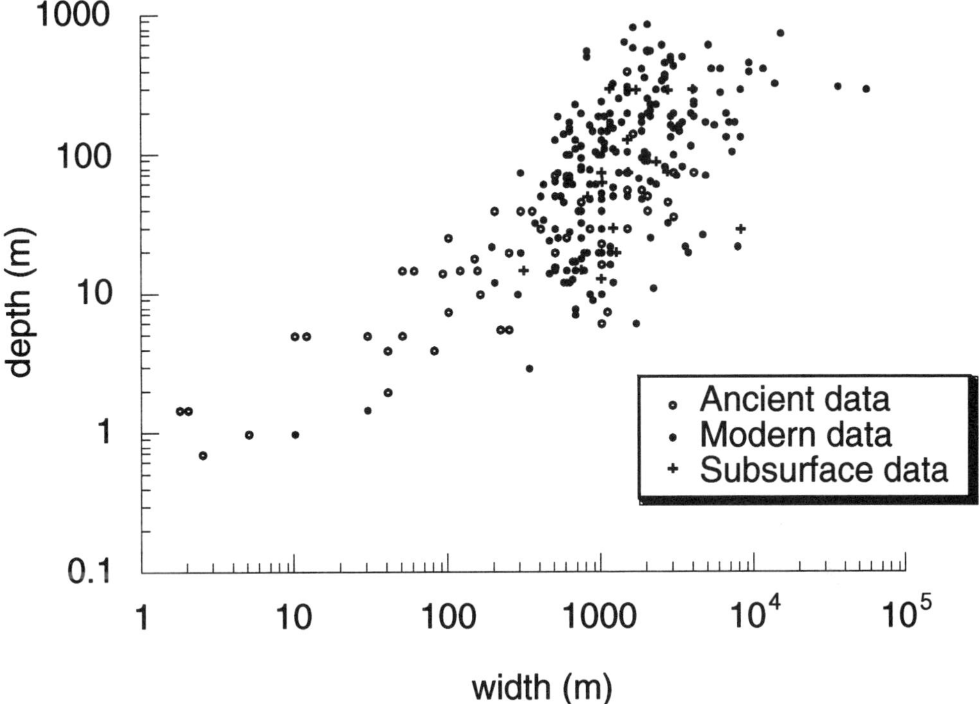

Figure 9.17. Plot of submarine channel width and depth data from various modern, ancient and subsurface channels.

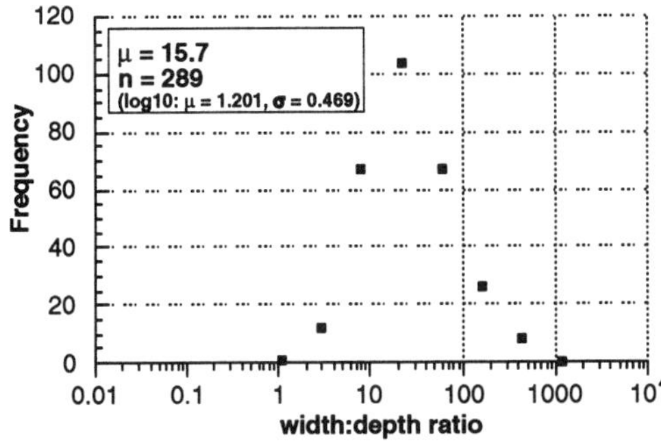

Figure 9.18. Graph to show the approximate log-normal distribution of submarine channel width:depth ratio data.

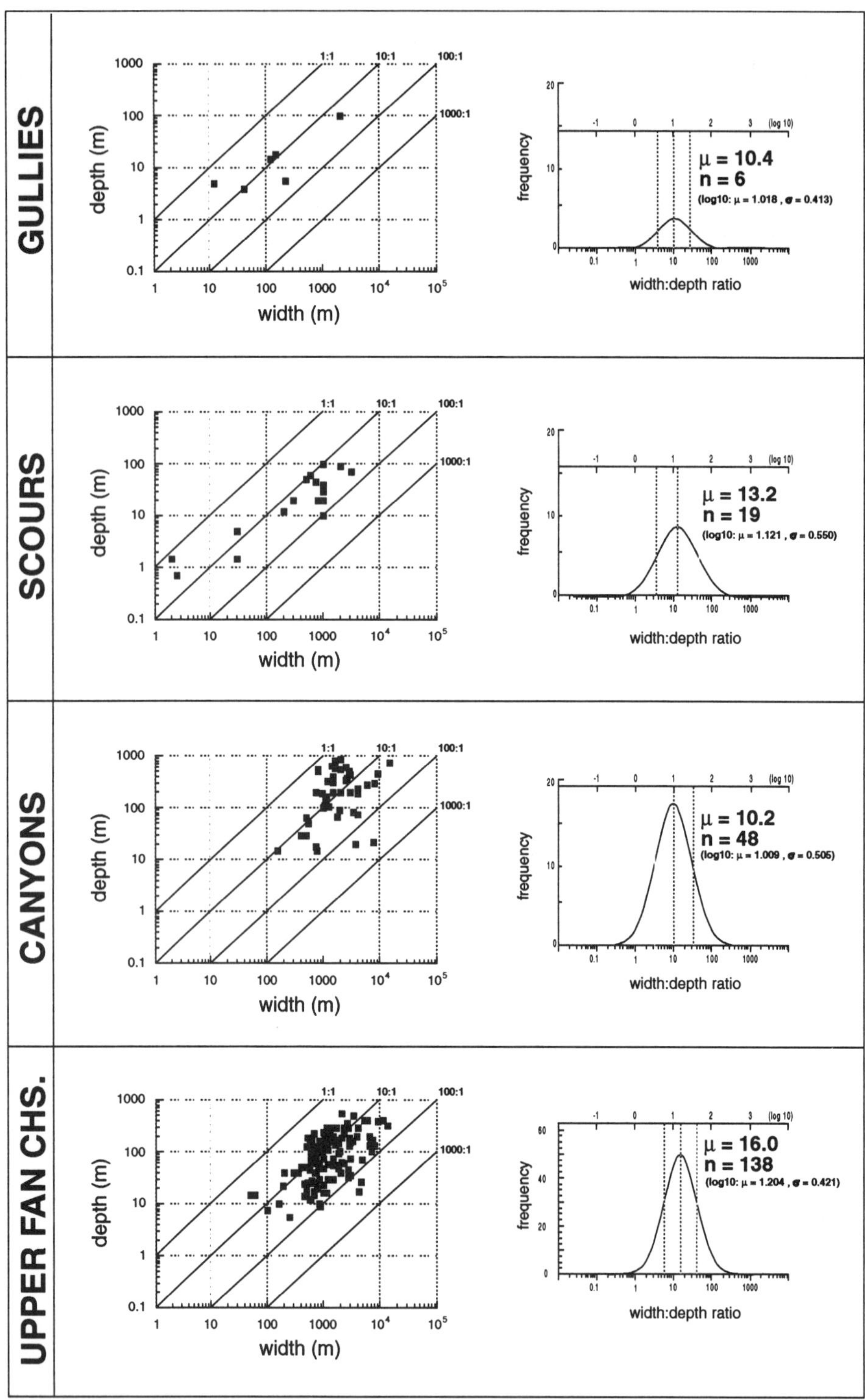

Figure 9.19. Graphs showing the aspect ratios of gullies, canyons, scours, upper fan channels, middle fan channels, lower fan channels and mid-ocean channels.

198

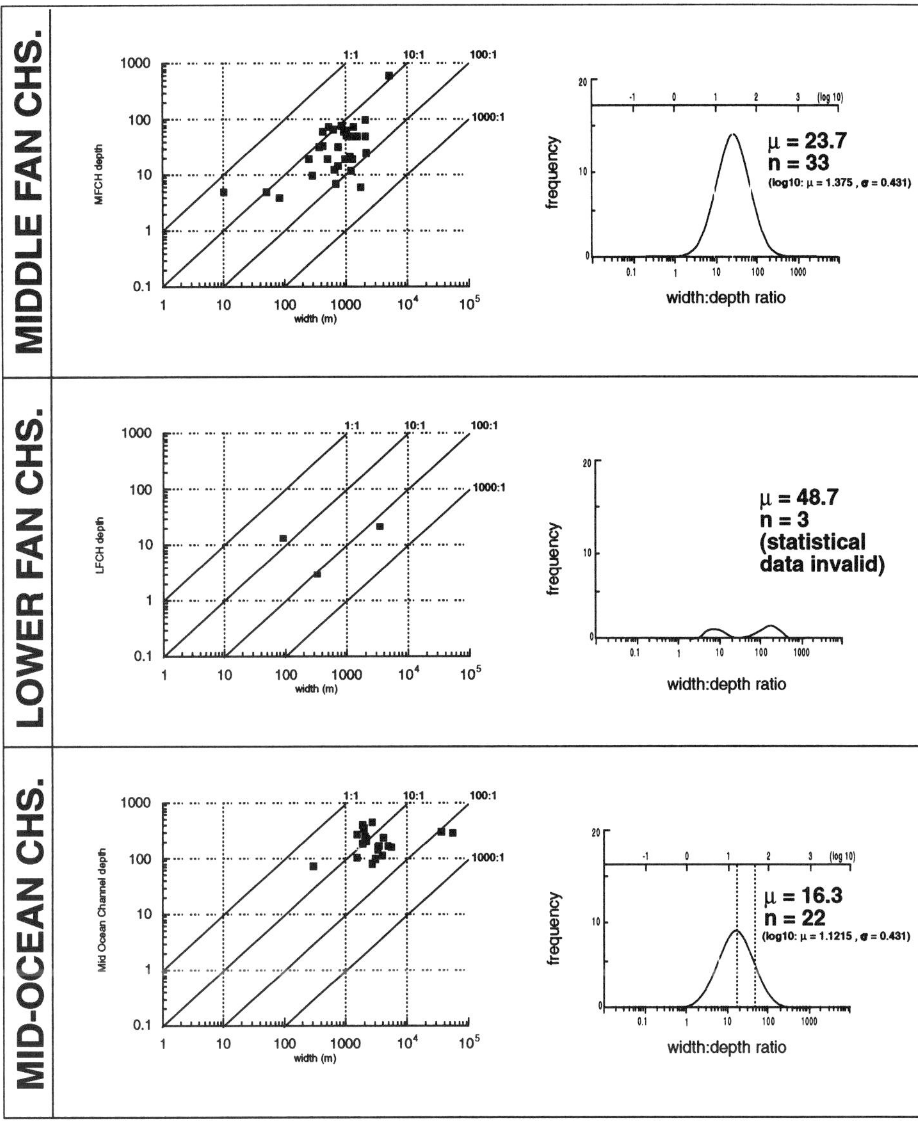

Figure 9.19. Continued.

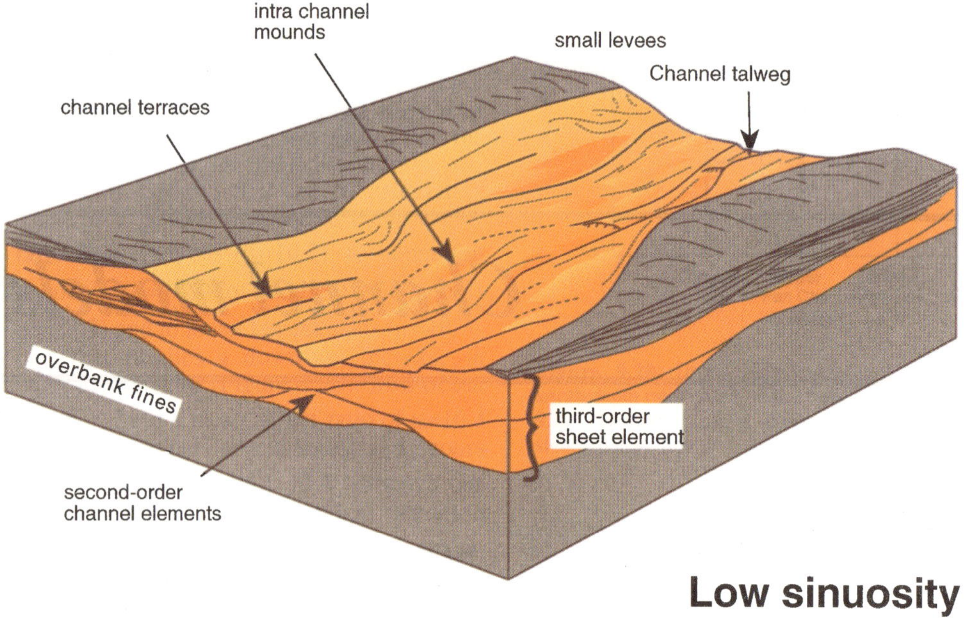

Figure 10.1. Architectural element model for continued erosional submarine channels. See text for details.

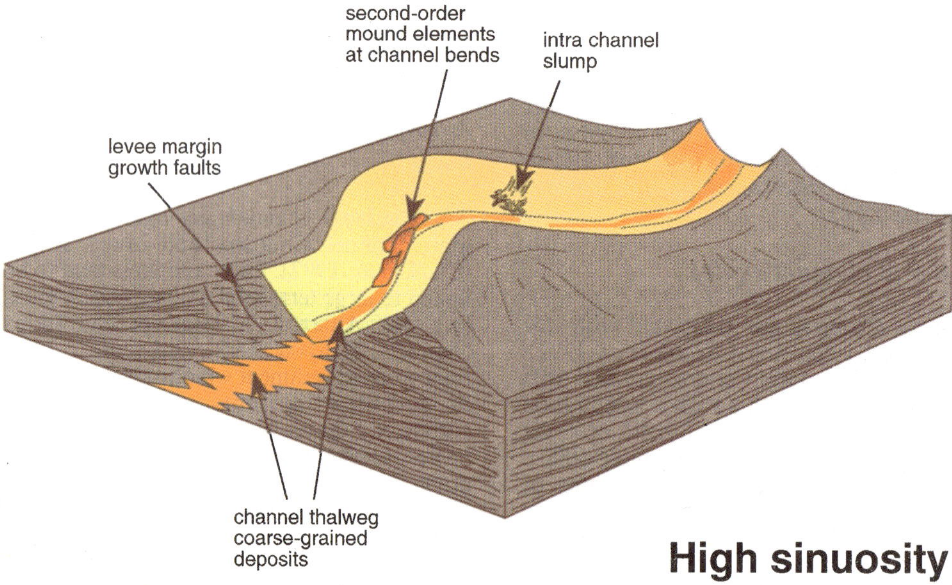

Figure 10.2. Architectural element model for continued aggradational (or depositional) submarine channels. See text for details.

material, channel growth is mainly vertical (e.g. the Indus channels). The aggradation of large channel-levee complexes however, commonly results in major avulsion from one site to another, resulting in discontinuous bodies of channel axis deposits.

Application of erosional channel model to ancient channel infills

As indicated above, in many documented ancient turbidite exposures, it is not possible to "walk-out" beds from within a channel into the levee or overbank deposits. In ancient outcrops, we suspect that one of the main reasons for encountering difficulties with tracing beds from intrachannel to extrachannel sites results from the recognition of only the lower parts of channel fills, and possibly even the mis-identification of parts of canyon-fills as middle-fan channels.

Channel depth estimates rarely include the thickness of much fine-grained fill, contrary to the observations in some modern fan channels (e.g. Mississippi Fan DSDP Leg 96 Shipboard Scientific Party 1983). In ancient outcrops, fine-grained lithologies (muds and silts) are generally poorly exposed at critical outcrops, and, or, may have undergone severe post-depositional deformation to obscure original bedding relationships. Such severe limitations frequently conspire to encourage an under-estimation of true channel dimensions and the interpretation of packets of fine-grained sediments more than metres in thickness as "levee", "interchannel", or "overbank" deposits. This means that the lower parts of many ancient channel fills, commonly recognised by large-scale erosional surfaces overlain by coarse-grained facies (Facies Classes A, B and C, Pickering *et al.* 1986b, 1989), should only show sharp contacts with the erosional surface and, as exemplified in the Ainsa I channel margin, classic onlap relationships (see Figure 7.23). In these and many other ancient examples, we suspect that much of the fine-grained facies represent mainly slope sediments sensu stricto into which the channels were incised, and that possibly large portions of ancient channels are infilled by muds and silts, not coarser sediments.

Sequence stratigraphy and architectural elements

The 1980s witnessed the development of the hypothesis of sequence stratigraphy, the progeny of seismic stratigraphy from the 1970s (Mitchum 1977, Mitchum *et al.* 1977). Useful definitions include "Sequence stratigraphy is the study of rock relationships within a chronostratigraphic framework of repetitive, genetically-linked strata bound by surfaces of erosion or non-deposition, or their correlative conformities" (van Wagoner *et al.* 1988), and "Sequence stratigraphy is the study of genetically related facies within a framework of chronostratigraphically significant surfaces. The sequence is the fundamental stratal unit for sequence stratigraphic analysis" (van Wagoner *et al.* 1990). Sequence stratigraphy provides a useful framework for understanding the interplay between accommodation space, subsidence and/or uplift, and rates of sea-level change in many sedimentary environments (e.g. van Wagoner *et al.* 1988, 1990).

Although philosophically similar in many respects to the Exxon approach, Galloway (1989a, 1989b) has put forward the concept of a genetic stratigraphic sequence. A genetic stratigraphic sequence is the sedimentary product of a depositional episode, each sequence comprising (1) a progradational facies-association; (2) an aggradational facies-association, and (3) a retrogradational or transgressive facies-association. Genetic sequences are bound by a sedimentary veneer or surface that records the depositional hiatus that occurs over much of the transgressed shelf and adjacent slope during maximum marine flooding.

Einsele (1985) has considered the response of sediments to sea-level changes in differing storm-dominated margins and epeiric seas, particularly the Mesozoic epicontinental, mud-dominated, seas in the slowly subsiding basins of Germany. Such basins contrast markedly with the rapidly subsiding shelf-margin seas subject to rapid changes in sea level that are glacio-eustatically driven, and from which much of the Exxon philosophy is based. He pointed out that the base level to which sediments aggrade is the storm wave base rather than sea level sensu stricto. Einsele (1985) believes that sediment accumulation patterns are particular to depositional sites and conditions.

Perhaps, the principal difference between the Exxon model and Einsele's (1985) perspective is that: (1) regressions are perceived as gradual rather than abrupt; (2) purely aggradation depositional units reach their maximum thickness where basin subsidence is most rapid, for example in the centre of basins, and (3) the base level to which sediments may aggrade is the storm wave base.

Although the global synchroneity, or correlatability, of events in relation to the proposed eustatic sea-level curves presented by Haq *et al.* (1977, 1988) has been undermined (see discussion by Miall 1992), the principles of sequence stratigraphy, the associated ways of dividing up stratigraphic units, and interpreting many of the causal mechanisms, remain valid. For these reasons, a thorough understanding of the basic principles of sequence stratigraphy are important, but at the present time their applicability in deep marine environments remains more problematical than is the case in shallow marine settings. For a useful review of the nature and causes of cyclicity in turbidite and related deep-water systems, the reader is referred to 'Cycles and Events in Stratigraphy' edited by Einsele *et al.* (1991).

Relative lowstands in sea level tend to be associated with lowstand turbidite systems, as basin-floor and/or slope fans, for example as has been demonstrated for the Paleogene of the Central and northern North Sea (Armentrout *et al.* 1993, Hartog Jager *et al.* 1993, Galloway *et al.* 1993, Vining *et al.* 1993). This is because a fall in relative sea level, which may be eustatically driven, leads to the effective narrowing of shallow-marine shelves and the propensity for fluvial/coastal systems to prograde towards, even reaching, the shelf-slope break. Additionally, under conditions of lowered sea level, shallow-marine processes, such as severe storm wave pounding, are more likely to enhance sediment failure along basin slopes and, therefore, further increase sediment delivery rates into deep-marine environments.

Tectonic and/or climatically driven changes in the source area, even in the absence of any significant changes in sea level relative to the coastline, can increase sediment delivery rates to the deep sea and, therefore, favour the growth of deep-marine clastic systems (e.g. Kolla & Macurda 1988). For example, Mississippi Fan sandy turbidite sedimentation, initiated during the falling and

204

maximum relative lowstand stages of sea level during the last glacio-eustatic cycle, continued into the Holocene mid to late sea-level rise at about 12,000-11,000 years ago, due to: (i) landward extension of the Mississippi Canyon into the mid-shelf water depths as sea level rose; (ii) increased glacial meltwater discharge and pebble to clay size sediment loads delivered directly to the canyon head during rising sea level; (iii) persistent interception of longshore drift by the canyon as it eroded headward; (iv) steep gradients at the canyon head favoured sediment failure as sediment slides, debris flows and turbidity currents, and (v) sediment bypass of coarse-grained material through the canyon into deep water with the absence of deltaic stratal patterns within the canyon (Kolla & Perlmutter 1993). Kolla and Perlmutter (1993) note that the late sand-prone turbidite sedimentation in the Mississippi Fan is compatible with the occurrence of sandy turbidites in the middle Amazon Fan subsequent to 13,285+650 years ago. In both the Mississippi and Amazon Fan, however, these late sand-prone phases of fan growth may coincide with at least the earliest part of the Younger Dryas global cooling and, therefore, represent renewed lowstand sand accumulation, an aspect requiring further evaluation. Furthermore, in a discussion of lowstand deep-water siliciclastic depositional systems, Kolla (1993) has shown that the terms "basin-floor fan" and "slope fan" are too restrictive both literally and conceptually to represent the many aspects of such systems. We therefore prefer the term "active fan growth phase" to lowstand system.

In order to better constrain the sediment types, grain sizes and rates of sediment accumulation, and the fundamental controls on deposition, future research must focus on providing high-resolution stratigraphy in areas where chronostratigraphic surfaces can be correlated from basin to shelf (e.g. Brushy Canyon Formation, west Texas). Chemical methods, including isotopic analysis, provide a major way forward, particularly where global and/or regional climate change has exerted a major influence on deposition.

With the above caveats in mind, and assuming that most ancient documented channel complexes are erosional and erosional-depositional (sensu Mutti and Normark 1987), the following general model for ancient channel infill is proposed (Figure 10.3).

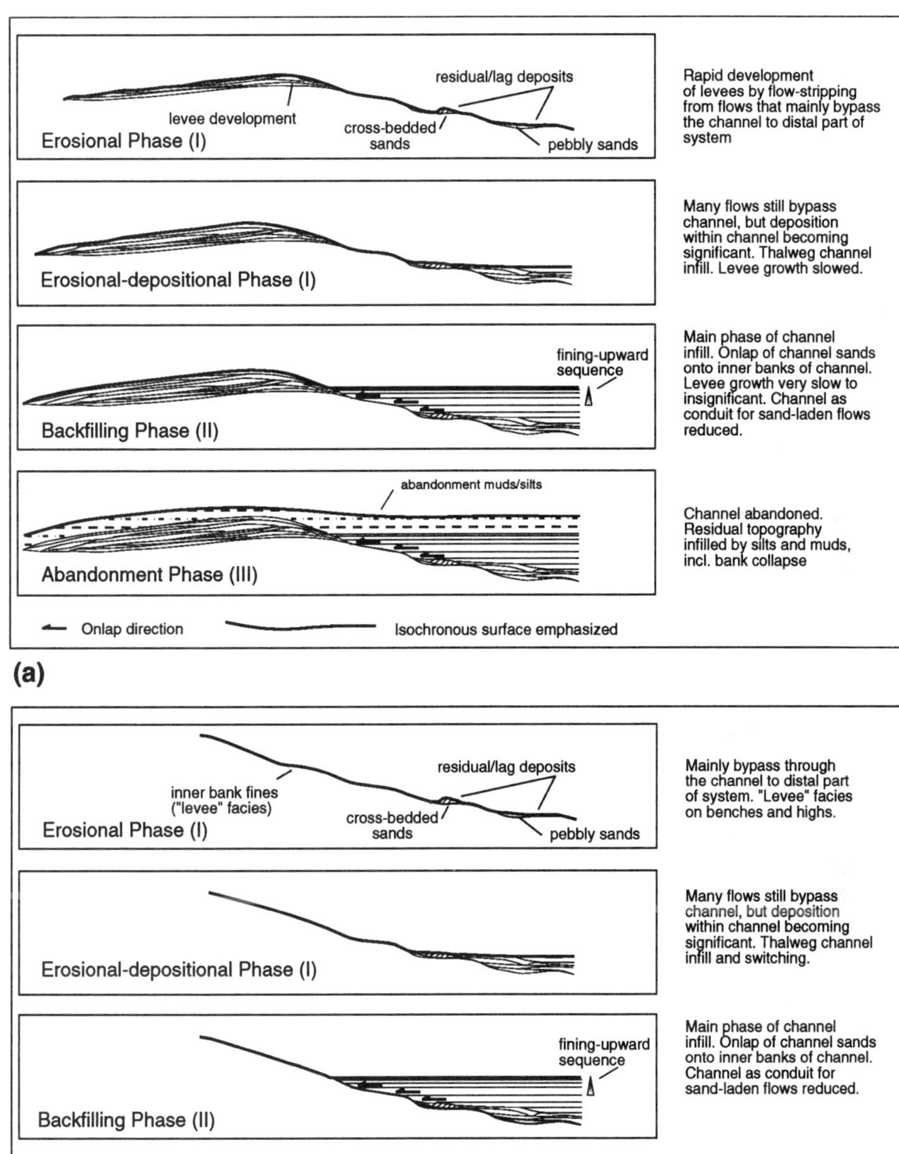

Figure 10.3. Generalized model for the excavation and infill of ancient deep-water channels, showing different phases of channel fill. (**a**) erosional/depositional channels, and (**b**) erosional channels, where the channel-fill deposit is entirely isolated from the channel levees. See text for explanation.

Phase I - Erosion, through-flow and non-deposition

The initial phase of channel development involves the excavation, commonly by multiple discrete erosional events, of a complex, commonly stepped or benched, erosional surface. During this phase, small relatively straight slope gullies may form on the slope, and grow both by headward erosion and an increase in sinuosity (*cf.* canyon development, Twichell and Roberts 1982, McGregor *et al.* 1982). The initial position of channels will be determined by many factors, both extra- and intra-basinal, the specifics of which must be determined individually for each channel complex. As the channels develop and extend in both a headward and basinward direction, the relatively straight courses will tend to become more sinuous, particularly where gradients are lower .

During Phase I, the channel acts essentially as a conduit for the through-put of channelised sediment gravity flows and is, therefore, the time of maximum development of channel lag deposits (e.g. as pebble stringers etceteras), reworking of previously deposited sands into bedform fields (e.g. cross-stratified sands), and irregular and localised scouring of the channel floor. Additionally, channel overspill and flow-stripping are most common during Phase I times, leading to maximum growth of the levees - because through-channel flows have their greatest competence and capacity compared to other times of channel development. Phase I is associated with the most irregular bed geometry, irrespective of the sediment facies.

Phase II - Deposition

This is the main phase of channel infill by coarse-grained facies, in which back-filling, sand-bar aggradation, point-bar growth (if relevant), and other depositional processes occur. It is during this phase that the erosional channel walls are typically onlapped by more sheet-like beds compared to Phase I deposition. Phase II is associated with the most commonly described relatively coarse-grained intrachannel facies.

Phase III - Abandonment

The transition from Phase II to Phase III deposition has the greatest propensity to generate thinning-and-fining-upward sequences, as the channel becomes abandoned as a conduit for sandy flows and their deposits. Channel abandonment may be caused by many processes, e.g. channel avulsion further up-system and the progressive plugging of the newly constructed channel margin-levee, leading to systematically less overspill of coarser sediments along the older conduit, or an extra-basinal change in base level, such as a relative rise in sea level or decreased tectonic activity in the source region. During Phase III the channel may be partially or completely filled by fine-grained facies.

Repetition of Phases I-III

It should be noted that any channel history may involve the repetition of any of the above phases, e.g. if the channel is reactivated because of changes in base level, it may go from Phase III to Phase I or Phase II. Consequently, the vertical and lateral facies relationships observed within channel fills are commonly both complex and unpredictable. For example, any vertical sequence through a channel fill may show several phases of deep scouring overlain by channel lag deposits, and it is at such times of channel reactivation that a channel may dramatically shift or migrate sideways. A good example of this process is seen in the Ainsa II channel complex, Eocene Hecho Supergroup, Spain, or the East Kongsfjorden Channel of the late Precambrian Kongsfjord Formation, northern Norway.

We would caution the correlation of basal major unconformities (e.g. type I unconformity) from shelf environments (e.g. the bounding unconformities to major valley incision) to the base of sand packets in deep-marine environments (*cf.* Mutti *et al.* 1989). The actual shelf-to-basin correlative surface within a turbidite sand packet will probably be a basal bed contact within that packet not at its base due to the time lag between maximum shelf (valley) incision and its effects in the basinal system. Thus many type I unconformities, unless specifically traced through the shelf-slope-basin system are likely to be miscorrelated. In the absence of better constraints, any sensible sequence stratigraphic approach to deep-marine systems should correlate major shelf valley incision with packets of beds in the basin and lower slope. In ancient outcrops, the base of many channels commonly is observed as less erosive than parts of the lower channel-fill events, probably reflecting initial channel excavation as base level falls (e.g. regression) but with maximum incision at some time later.

Model for channel evolution in moderate to high sinuosity channels.

In many well-documented ancient channels systems that are interpreted as moderate- to high-sinuosity channels (based on the recognition of lateral accretion elements and from 3-D seismic mapping), it appears that there may be a predictable evolution from the initial erosional to the final abandonment phase. Here, we propose a new model (Figure 10.4) that needs testing.

The establishment of a channel involves an initial erosional phase associated with the deposition of mainly residual deposits and irregular bedding/elements (stage 1 deposits). In many cases, the channel-floor topography probably develops from a number of discrete erosional events in which channel terraces may be formed.

Channels will tend towards an equilibrium profile by becoming more sinuous, unless there are modifying effects that lead to abrupt abandonment of the channel or lead to re-activation of the initial erosional phase (e.g. caused by base-level changes leading to increased rates of sediment supply in flows with greater competence and capacity). During the development of the sinuosity lateral accretion and flow-stripping processes (stage 2 deposits) will become significant in the resulting deposits and channel architecture. As the channel changes towards becoming abandoned, e.g. as the sediment flux decreases in flows with, on average, decreased competence and capacity, then the channel will tend to back-fill and change from offset-stacked inclined macroforms (due to lateral accretion and flow-stripping, or other channel bar forms) to lens elements (stage 3 deposits), and finally abandonment (stage 4 deposits) (Figure 10.4).

We would predict that levee bank collapse will tend to be most prevalent between stages 1 and 2 deposits as this represents the time when the levee-crest to channel-floor height is greatest and the levees are most unstable. Levees will show their greatest rate of vertical aggradation during the accumulation of stage 1 deposits since bankfull flow conditions are most likely during the passage of high-competence and high-capacity turbidity current flow.

The evolutionary model proposed here represents an idealised erosional-depositional history and contains a number of implicit assumptions, including: (i) a gradual change between different typical flow conditions within any reach of a channel; (ii) the absence of any major sediment slide/debris-flow event that leave an anomalously thick deposit within a channel to plug, or partially plug, a reach of the channel and cause channel avulsion, and (iii) that the channel is excavated, acts as a conduit for sediment gravity flows, and is filled to abandonment, before being over-supplied with relatively coarse-grained clastics, all within one cycle of falling then rising base level such as a falling then rising sea-level curve. Naturally, there are many exceptions to this model but we are encouraged by the observation that many ancient channel systems appear to conform to this evolutionary history (e.g. Eocene Ainsa I, II and Gerbe II channels, Permian Indian Draw Field etc.).

Post-depositional modification of channel fill

Post-depositional modification of channel deposits exerts a primary influence on the observable features and characteristics of ancient channels, and can control the location of subsequent channels in active deep-marine systems. Post-depositional modification features include clastic dykes and sills (including the formation of brecciated beds), faults (including growth faults), and differential compaction.

Fluidisation and liquefaction features, such as clastic dykes and sills are documented from many ancient, including industry subsurface, deep-water systems, e.g. associated with slope sandstone gully fills (Surlyk 1987), inferred submarine channel fills (Hiscott 1979, Pickering 1981, 1983). In the subsurface, good examples of large-scale sand-rich injection structures in the Paleogene of the northern North Sea include sand dykes up to many metres wide intruding vertically up through tens of metres of sediments in the Balder Formation, as described by Jenssen *et al.* (1993), and on a decimetre scale in core from the Gryphon Field (Newman *et al.* 1993).

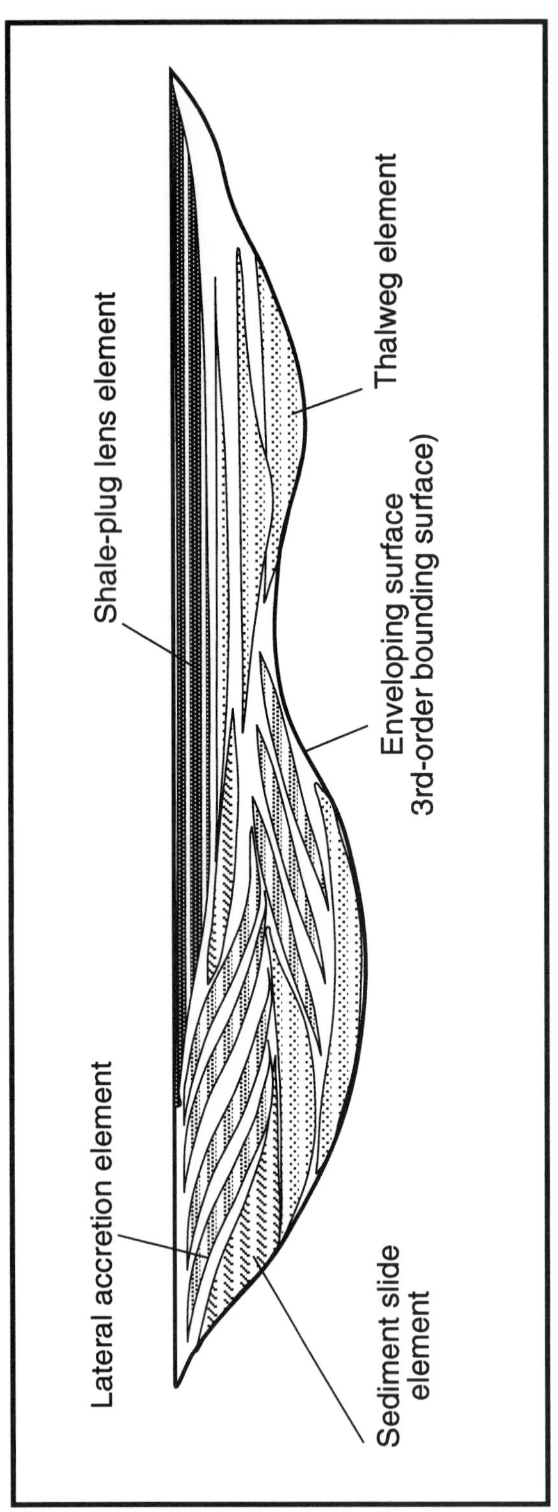

Figure 10.4. Schematic model for the development of architectural elements within mixed erosional/depositional channel systems. The development of channel sinuosity during the infilling phase of channel development results in the deposition of lateral accretion and flow-stripping elements (stage 2 deposits), above the initial "residual" elements (stage 1 deposits) resulting from the erosion phase. Stage 3 deposits will typically be lens elements resulting from channel "back-filling", and finer-grained abandonment facies constitute the final part of the channel-fill (stage 4 deposits).

A common feature recorded in many industry subsurface cores, e.g. from the Paleogene of the northern North Sea, is the presence of apparently chaotic mud-flake breccias and conglomerates with a sandy matrix. Although, on the basis of micropalaeontology and composition, including colour, some of these deposits are interpreted as sediment slides and debris flow deposits (with extra- and intra-formational clasts), there are many cases where it appears that the deposit was formed by the pervasive injection of turbiditic and hemipelagic muds. Examples occur where the lower part of a sandy turbidite bed has fluidised and pervasively injected the upper muddy part of the same bed, a process here referred to as "autobrecciation".

Growth faulting, particularly at channel margins associated with differential compaction, may encourage the offset-stacking of channel sand bodies. Growth faults also may act as surfaces and zones of weakness along which large-scale wet-sediment intrusions occur. Growth faults may be reactivated with compaction, even very early after deposition when only perhaps tens of metres of additional stratigraphy has accumulated. The location of compaction-related faults can be controlled by buried channel courses, paralleling the channel margins, caused by the differential compaction of channel sands relative to levee and overbank finer grained sediments (see fig. 4 in Jenssen *et al.* 1993, fig. 8 in Newman *et al.* 1993).

Differential compaction of ribbon-like channel sand bodies and their associated overbank-levee deposits can lead to inversion of the primary topographic relief as the sands compact less than the enveloping muds. Such post-depositional processes may be associated with the development of growth faults along the margins of the buried channels, and even the injection of sandy clastic dykes along such faults.

Chapter 11

Hydrocarbon prospectivity

Submarine channel can act as productive hydrocarbon intervals (e.g. Hartog Jager *et al.* 1993, Newman *et al.* 1993, Timbrell 1993, Vining *et al.* 1993). Coarse-grained turbidites commonly have high porosity-permeability values, but a common problem in reservoir engineering is the evaluation of the effects of the turbidite mud and shale caps on vertical permeability, K_v. Increased amalgamation of turbidites beds towards the axis of channels, relative to the channel margins, would be expected to increase sandstone bed connectivity and therefore result in an increase in the K_v/K_h ratio from the channel margins towards the axis. The stacking pattern and interconnectivity of sand-rich channel elements, otherwise enclosed by permeability barriers, is discussed below.

Reservoir geometry

In order to better understand the heterogeneity within channel reservoirs, it is necessary to have a detailed knowledge of the different architectural elements. Table 11.1 lists some of the architectural elements discussed, in addition to a selection of channel elements taken from published literature, to show their dimensions and the degree of architectural element lateral continuity and vertical connectivity. Lateral continuity is defined as the width:depth ratio measured perpendicular to flow/palaeoflow direction, while vertical connectivity is an estimate of the fraction of the element perimeter which is in contact with a similar element (Figure 11.1) (see Fielding & Crane

1987). The lateral continuity and vertical connectivity of architectural elements are graphically shown in Figure 11.2 which also shows the cross-sectional area of the individual elements. Such a graphical representation of channel elements is useful in the quantitative comparison of submarine channels and channel elements. Not all the elements plotted on Figure 11.2 can be described as reservoir or reservoir analogue bodies, for example large mud-rich channel-levee architectural elements of the Amazon, Mississippi and Indus fans. Nevertheless, most of the examples that have been used represent candidate reservoir analogue bodies and, as such, can be useful in establishing quantitative analogue models.

Preferential sites of sand accumulation in channels

The physical processes operating within submarine channels, particularly associated with turbidity current flow, appear to result in at least some predictability about sites of preferential sand accumulation and, therefore, potential hydrocarbon stratigraphic traps. Flow stripping at channel bends, and probably channel confluences, should result in the deposition of sand lenses towards the outer channel margin. Continued flow stripping may produce offset-stacked sand lenses. In contrast, where channel lateral migration is important, as in high-sinuosity channel systems, sand accumulation is predicted to occur on the inner channel bank to create lateral accretion surfaces. At present, there is insufficient well-constrained

ELEMENT	width (m)	depth (m)	lateral continuity	vertical connectivity	C.S.A. (m^2)	reference
Ainsa I "back-fill"" element beds"	200	3	65	0.2	6.0×10^2	1
Ainsa I channel	850	30	25	0	2.6×10^4	1
Ainsa I thalweg	230	4	60	0.4	9.2×10^2	1
Ainsa II channels	600	25	24	0.5	1.5×10^4	2
Ainsa II/2 channel/lens elements	50	2	25	0.35	1.0×10^1	2
Ainsa II/3 channel/lens elements	100	3	33	0.2	3.0×10^2	2
Ainsa II/4 axial fill element beds	100	2	50	0.9	2.0×10^2	2
Ainsa II/5 slide elements	100	4	25	0	4.0×10^2	2
Almeria Channel	400	50	8	0	2.0×10^4	3
Almeria Channel thalweg	65	5	13	0	3.3×10^2	3
Amazon-Middle Fan ch.-axis complexes	1250	150	8	0	1.9×10^5	4
Amazon-Middle Fan ch.-levee complexes	30000	150	200	0.9	4.5×10^6	4
Balder Fm. stacked channels	800	50	16	0.1	4.0×10^4	5
Black Flysch scours	30	5	6	0.2	1.5×10^2	6
Brushy Canyon Fm. "Salt-flat" Channel	2000	30	66	0	6.0×10^4	7
Brushy Canyon Fm. "100-foot" Channel	400	30	13	0	1.2×10^4	7
Brushy Canyon Fm. Popo channels	160	10	16	0.4	1.6×10^3	7
Brushy Canyon Fm. Brushy Mesa elements	155	15	10	1	2.3×10^3	7
Caban Coch channels	4000	75	50	0.2	3.0×10^5	8
Capistrano channels	250	20	12.5	0.4	5.0×10^3	9
Capistrano lat. accr. elements	50	2	25	1	1.0×10^2	9
Capistrano sheet-fill elements	100	1	100	1	1.0×10^2	9
Hamningberg Channel axial fill	100	10	10	0.3	1.0×10^3	10
Hamningberg channel margin element	75	5	15	0.3	3.8×10^2	10
Indian Draw Field A1 Sandstone elements	400	3	133	0	1.2×10^3	11
Indian Draw Field A3 Sandstone elements	500	5	100	0	2.5×10^3	11
Indian Draw Field channel-fill	1200	30	40	0	3.6×10^4	11
Indus-Middle Fan ch.-axis complexes	1250	75	17	0	9.4×10^4	12
Indus-Middle Fan ch.-axis elements	1000	15	66	0.9	1.5×10^4	12
Indus-Middle Fan ch.-levee complexes	25000	75	333	0.9	1.9×10^6	12
Indus-Upper Fan ch.-axis complexes	10000	400	25	0	4.0×10^6	12
Indus-Upper Fan ch.-axis elements	1500	20	75	0.9	3.0×10^4	12
Indus-Upper Fan ch.-levee complexes	50000	400	125	0.9	2.0×10^7	12
Indus-Upper Fan Channel C axis elements	5000	100	50	0.6	5.0×10^5	13
Indus-Upper Fan Channel Ca axis elements	3000	100	30	0.7	3.0×10^5	13
Indus Upper-Fan Channel Cc axis elements	2500	125	20	0	3.1×10^5	13
Milliners Arm Fm. axial elements	500	20	25	0.3	1.0×10^4	14
Milliners Arm Fm. ch. margin elements	750	15	50	0.4	1.1×10^4	14
Mississippi levees	600	10	60	0.9	6.0×10^3	15
Mississippi migrating thalweg	2000	100	25	0.4	2.0×10^5	15
Montagne de Chalufy turbidite channel	500	50	10	0	2.5×10^4	16
Monterey channel	1500	50	30	0	7.5×10^4	17
Rapitan Channel (RCH-1)	1500	75	20	0.5	1.1×10^5	18
Rapitan channel-fill elements (RCH-2a)	1200	15	80	0.2	1.8×10^4	18
Rapitan channel-fill elements (RCH-3a)	1000	8	125	0.05	8.0×10^3	18
Rhone-channel thalweg	500	130	4	0	6.5×10^4	19
Rhone-Neofan scours	1000	20	50	0.1	2.0×10^4	20
Rhone-Upper Fan leveed valley	1500	100	15	0	1.5×10^5	19
Risfjord channel stacked channels	50	5	10	0.9	2.5×10^2	10
Solitary Channel	200	40	5	0	8.0×10^3	1
Solitary Channel bedforms	150	4	37.5	0.25	6.0×10^2	1

Table 11.1. Dimensions and degree of lateral continuity and vertical connectivity of a selection of channel elements from published literature, including those discussed in previous chapters. Dimensions are measured perpendicular to flow/palaeoflow direction with the exception of the Tabernas Basin Solitary Channel bedforms which are measured in a down-current direction. 1 - author's own work; 2 - Clark (1995); 3 - Cronin et al. (1995); 4 - Damuth et al. (1995); 5 - Timbrell (1993); 6 - Vicente Bravo & Robles (1995); 7 - Zelt & Rossen (1995); 8 - Smith et al. (1991); 9 - Walker (1975c); 10 - Pickering (1982a); 11- Philips (1987); 12 - Kenyon et al. (1995a); 13 - McHargue (1991); 14 - Watson (1981); 15 - Pickering et al. (1986b); 16 - Hilton & Pickering (1995); 17 - Masson et al. (1995); 18 - Remacha et al. (1995); 19 - O'Connell et al. (1991); 20 - Kenyon et al. (1995b).

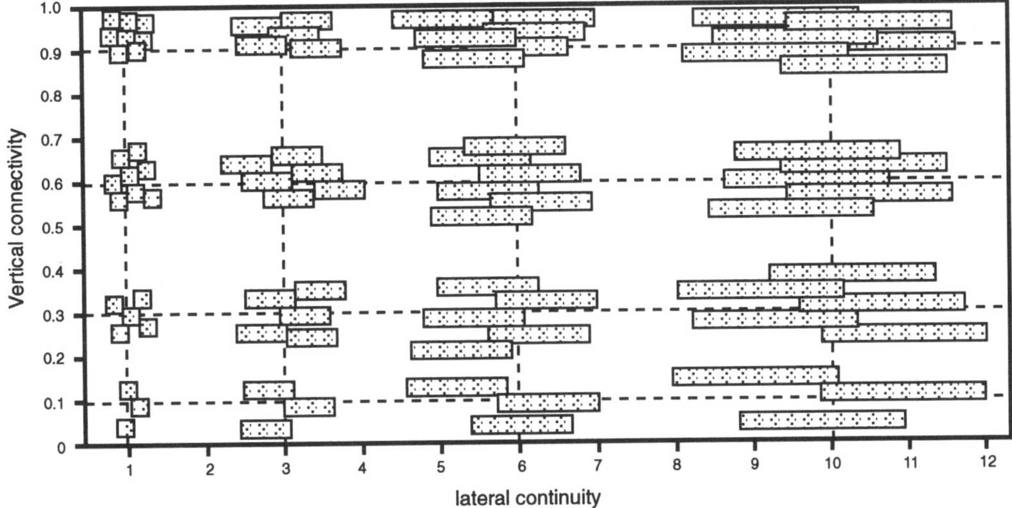

Figure 11.1. Schematic graph to show how architectural elements may form individual or composite bodies, and how the lateral continuity and vertical connectivity of these bodies determines reservoir size. Lateral continuity is defined as the width:depth ratio measured perpendicular to flow/palaeoflow direction, and vertical connectivity is an estimate of the fraction of the element perimeter area which is in contact with a similar element.

detailed 3D data available to evaluate the relative importance of flow-stripping versus lateral accretion as processes leading to preferential sites for sand accumulation in high-sinuosity channels. Additionally, the importance of apparently random sand deposition, probably due mainly to source-controlled variations in flow competence and capacity, and sediment calibre, requires evaluation.

Low-sinuosity deep-water channels appear to be more characteristic of relatively high gradient slopes, particularly off narrow shelves in tectonically active areas, e.g. California Borderland, and the Paleogene of the northern North Sea. Although in such settings flow-stripping is still an important process in controlling sites of preferential sand accumulation, especially where seafloor topography is significantly modified by tectonic features, the inherently more complex nature of such systems makes any model-based predictions about sand deposition considerably more difficult. Low-sinuosity channel systems appear to be characterised by channel benches and terraces, which are good candidate sites for turbidity currents to undergo successive hydraulic jumps and, therefore, cause the preferential deposition of sand. The high-resolution TOBI data from the Monterey channel appears to show high-backscatter facies bedform fields immediately down-channel from transverse benches. Whilst modern seafloor topography, sidescan sonar mapping, and bottom photography, ideally with sediment recovery, may permit a good understanding of the position of hydraulic jumps, such data are less readily available in the ancient rock record and, therefore, subject to a much greater measure of uncertainty.

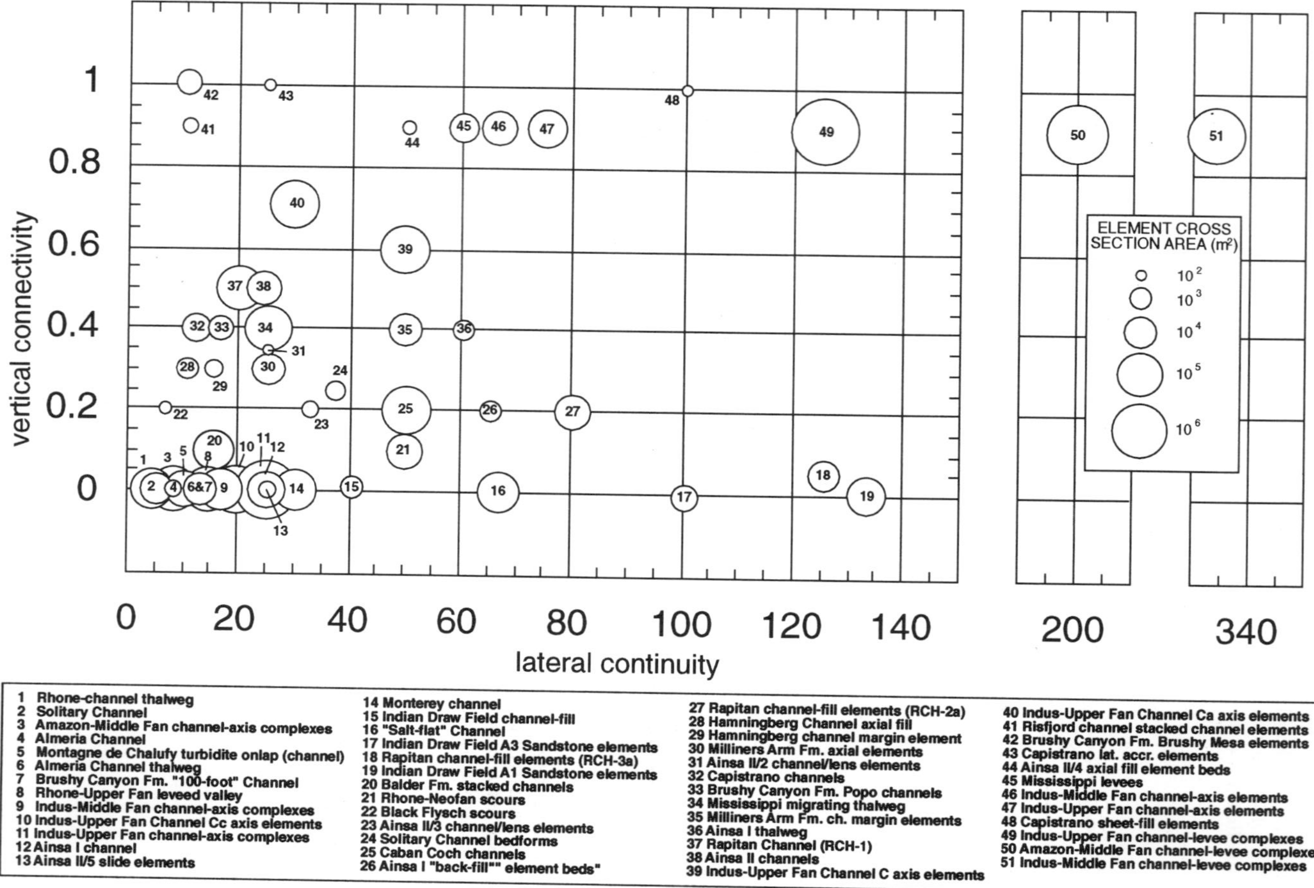

Figure 11.2. Graph showing lateral continuity and vertical connectivity of architectural elements and cross-sectional area of individual elements listed in Table 11.1. This approach is useful for comparison of reservoir bodies and reservoir analogues, although not all the elements plotted could be described as reservoir bodies (e.g., large mud-rich channel-levee architectural elements of the Amazon, Mississippi and Indus fans).

Clearly architectural element analysis of channels and their internal elements provides a useful tool in unravelling the complexities of different sedimentological stages of development. The breakdown of channel-fill sequences, and channel features, into their constituent architectural elements, leads to interpretations into the channel-fill type, the interconnectivity of channel bodies resulting from channel stacking, causal sedimentary processes, and architectural relationships between channels and levees, growth patterns, and channel stacking architecture. Predictions for the location of specific architectural elements within channels, or parts of channels, can be made, from inferred channel processes. In summary sites of preferential sand accumulation in submarine channels can be listed as follows:

- Channel bends associated with flow stripping.
- Channel confluences.
- Point bars associated with high-sinuosity lateral accretion.
- Channel benches and terraces.
- Channel thalwegs.
- Intrachannel hydraulic jump sites, e.g. transverse channel benches caused by deep erosional features and growth faults.

Comparisons between modern and ancient channel systems requires the identification of similar channel and intra-channel architectural elements. Suites of architectural elements can be characterised as erosional, erosional-depositional, or depositional, leading to the identification of channel type. This architectural element characterisation of submarine channels is very useful for comparative analogies between modern and ancient channels.

Many researchers have encountered discrepancies between modern and ancient channel architecture. One reason why such discrepancies should exist is that most ancient deep-water systems record deposition in small upper-slope, intra-shelf, and, or, aulacogen basins. These basins typically have high gradients compared to deep-water systems found on continental margins and oceanic basins. Slope gradients strongly control submarine channel processes, and steep slopes generally lead to the development of low-sinuosity channels (Clark *et al.* 1992).

A generalised architectural model for erosional and erosional-depositional channels (such as are common to the majority of ancient channel sequences) has been developed from architectural element analysis of submarine channel features. A phase of erosion, through-flow, and non-deposition occurs first, during which channel overspill and flow-stripping processes are most common, leading to the maximum aggradational growth of channel levees. This is followed by a phase of channel deposition, where backfilling and aggradational processes result in major infilling of the channel axis. An abandonment phase results in the final infilling of the channel. These phases of deposition may be repeated throughout the channel's growth due to reactivation and renewed phases of incision of channel systems resulting from changes in accommodation space in the source region caused by changes in sediment supply, tectonic activity and, or, changes in regional or global sea level.

References

ADAMS, J. 1990. Paleoseismicity of the Cascadia Subduction Zone: evidence from turbidites off the Oregon-Washington margin. *Tectonics*, **9**, 569-583.

ALLEN, J.R.L. 1983. Studies in fluviatile sedimentation: bars, bar-complexes and sandstone sheets (low-sinuosity braided streams) in the Brownstones (L. Devonian), Welsh Borders. *Sedimentary Geology*, **33**, 237-293.

ARMENTROUT, J.M., MALECEK, S.J., FEARN, L.B., SHEPPARD, C.E., NAYLOR, P.H., MILES, A.W., DESMARAIS, R.J. & DUNAY, R.E. 1993. Log-motif analysis of Paleogene depositional systems tracts, Central and Northern North Sea: defined by sequence stratigraphic analysis. *In*: PARKER, J.R. (ed.), *Petroleum Geology of Northwest Europe: Proceedings of the 4th Conference*, 45-57. Bath: The Geological Society.

BAGNOLD, R.A. 1962. Auto-suspension of transported sediment: turbidity currents. *Proceedings of The Royal Society of London*, A-**265**, 315-319.

BARNARD, W.D. 1978. The Washington continental slope; Quaternary tectonics and sedimentation. *Marine Geology*, **27**, 79-114.

BELDERSON, R.H., KENYON, N.H., STRIDE, A.H. & STUBBS, A.R. 1972. *Sonographs of the sea floor*. Amsterdam: Elsevier Publishing Company.

BELLAICHE, G., ORSOLINI, P., PETIT-PERRIN, B., BERTHON, J.L., RAVENNE, C., COUTELLIER, V., DROZ, L., ALOISI, J., GOT, H., MEAR, Y., MONACO, A., AUZENDE, J.M., BEUZART, P. & MONTI, S. 1983. Morphology au Sea-Beam de l'éventail sous-marin profond du Rhône et de son canyon afférent. *Comptes Rendus L'académie des Sciences Paris*, **296**, 579-583.

BELLAICHE, G., COUTELLIER, V. & DROZ, L. 1986a. Seismic evidence of widespread mass transport deposits in the Rhone deep-sea fan: their role in the fan construction. *Marine Geology*, **71**, 327-340.

BELLAICHE, G., COUTELLIER, V., DROZ, L. & MASSON, PH. 1986b. Deep-sea and Martian channels. *Deep-Sea Research*, **33**, 973-980.

BERING SEA EEZ-SCAN SCIENTIFIC STAFF, 1991. *Atlas of the US Exclusive Economic Zone, Bearing Sea*. US Geological Survey Miscellaneous Investigations Series I-2053, scale 1:500,000.

BOUMA, A.H., STELTING, C.E. & COLEMAN, J.M. 1984. Mississippi Fan: internal structure and depositional processes. *Geo-Marine Letters*, **3**, 147-154.

BOUMA, A.H., BARNES, N.E. & NORMARK, W.R. 1985a. *Submarine fans and related turbidite systems*. New York: Springer-Verlag.

BOUMA, A.H., COLEMAN, J.M. & DSDP LEG 96 SHIPBOARD SCIENTISTS 1985b. Mississippi Fan: Leg 96 program and principal results. *In:* BOUMA, A.H. BARNES, N.E. & NORMARK, W.R. (eds), *Submarine fans and related turbidite systems*, 247-252. New York: Springer-Verlag.

BOUMA, A.H., COLEMAN, J.M., ET AL. 1986. *Initial reports of the Deep Sea Drilling Project*, **96**. Washington, D.C.: US Government Printing Office.

BRIDGE, J.S. & LEEDER, M.R. 1979. A simulation model of alluvial stratigraphy. *International Association of Sedimentologists*, **26**, 617-644.

BRISTOW, C.S. & BEST, J.L. 1993. Braided rivers: perspectives and problems. *In:* BEST, J.L. & BRISTOW, C.S. (eds), *Braided rivers*. 1-11. Geological Society, London, Special Publication, **75**. Bath: The Geological Society.

BROOKFIELD, M.E. 1977. The origin of bounding surfaces in ancient aeolian sandstones. *Sedimentology*, **24**, 303-332.

BRYANT, W.R. & DSDP LEG 96 SHIPBOARD SCIENTISTS 1985. Consolidated characteristics and excess pore water pressures of Mississippi Fan sediments. *In*: BOUMA, A.H., BARNES, N.E. & NORMARK, W.R. (eds), *Submarine fans and related turbidite systems*, 299-309. New York: Springer-Verlag.

CARLSON, P.R. & NELSON, C.H. 1969. Sediments and sedimentary structures of the Astoria submarine canyon-fan system, NE Pacific. *Journal sedimentary Petrology*, **37**, 1269-1282.

CARTER, R.M. 1988. The nature and evolution of deep-sea channel systems. *Basin Research*, **1**, 41-54.

CARTER, R.M. & LINDQVIST, J.K. 1975. Sealers Bay submarine fan complex, Oligocene, southern New Zealand. *Sedimentology*, **22**, 465-483.

CARTER, R.M. & NORRIS, R.J. 1977. Redeposited conglomerates in Miocene flysch sequence at Blackmount, western Southland, New Zealand. *Sedimentary Geology*, **18**, 289-319.

CAZZOLA, C., FONNESU, F., MUTTI, E., RAMPONE, G., SONNINO, M. & VIGNA, B. 1981. Geometry and facies of small, fault-controlled deep-sea fan systems in a transgressive depositional setting (Tertiary Piedmont Basin, northwestern Italy). *In*: RICCI LUCCHI, F. (ed.), *Excursion Guidebook*, 5-56. 2nd Regional Meeting of International Association of Sedimentologists, Bologna.

CLARK, J.D. 1994. *Architecture and processes in modern and ancient deep-marine channel complexes*. Unpublished Ph.D. Thesis, University of Leicester, UK.

CLARK, J.D. 1995. A detailed section across the Ainsa II channel complex, south-central Pyrenees, Spain. *In*: PICKERING, K.T., HISCOTT, R.N., KENYON, N.H., RICCI LUCCHI, F. & SMITH, R.D.A. (eds), *Atlas of Deep Water Environments: Architectural Style in Turbidite Systems*, 139-144. London: Chapman & Hall.

CLARK, J.D. & PICKERING, K.T. 1996. Architectural elements and growth patterns of submarine channels: Application to hydrocarbon prospectivity. *American Association of Petroleum Geologists Bulletin*, **80**, 194-221.

CLARK, J.D., KENYON, N.H. & PICKERING, K.T. 1992. Quantitative analysis of the geometry of submarine channels: implications for the classification of submarine fans. *Geology*, **20**, 633-636.

CLIFTON, H.E. 1984. Sedimentation units in stratified deep-water conglomerates, Paleocene submarine canyon fill, Point Lobos, California. *In*: KOSTER, E.H.& STEELE, R.J. (eds), *Sedimentology of*

gravels and conglomerates, 429-441. Canadian Society of Petroleum Geologists Memoir, **10**.

COLEMAN, J.M., PRIOR, D.B. & LINDSAY, J.F. 1983. Deltaic influences on shelf edge instability processes. *In:* STANLEY D.J. & MOORE G.T. (eds), *The shelf break, critical interface on continental margins*, 121-137. Society of Economic Paleontologists and Mineralogists Special Publication, **33**. Tulsa, Oklahoma: Society of Economic Paleontologists and Mineralogists.

COLLINSON, J.D. 1968. The sedimentology of the Grindslow Shales and the Kinderscout Grit: a deltaic complex of the Namurian of northern England. *Journal of Sedimentary Geology*, **39**, 194-221

COLLINSON, J.D. 1970. Deep channels, massive beds and turbidity current genesis in the central Pennine Basin. *Proceedings of the Yorkshire Geological Society*, **37**, 495-519.

CRONIN, B.T. 1994. *Channel-fill architecture in deep-water sequences: variability, quantification and applications*. Unpublished Ph.D. Thesis, University of Wales, UK.

CRONIN, B.T. 1995. Structurally-controlled deep-sea channel course: examples from the Miocene of southeast Spain and the Alboran Sea, southwest Mediterranean. *In*: HARTLEY, A.J. & PROSSER, D.J. (eds), *Characterization of deep marine clastic systems*, 115-135. Geological Society, London, Special Publication, **94**. Bath: The Geological Society.

CURRAY, J.R. & MOORE, D.G. 1971. Growth of the Bengal Deep-Sea Fan and denudation in the Himalayas. *Bulletin of the Geological Society of America*, **82**, 563-572.

DAMUTH, J.E. 1980. Use of high frequency (3.5 kHz) echograms in the study of near-bottom sedimentation processes in the deep-sea; a review. *Marine Geology*, **38**, 51-75.

DAMUTH, J.E. & KUMAR, N. 1975a. Amazon Cone: morphology, sediments, age, and growth pattern. *Bulletin of the Geological Society of America*, **86**, 863-878.

DAMUTH, J.E. & KUMAR, N. 1975b. Late Quaternary depositional processes on the continental rise of the western equatorial Atlantic: comparison with the western North Atlantic and implications for reservoir rock distribution. *American Association of Petroleum Geologists Bulletin*, **59**, 2171-2181.

DAMUTH, J.E. & EMBLEY R.W. 1981. Mass-transport processes on Amazon Cone: western equatorial Atlantic. *American Association of Petroleum Geologists Bulletin*, **65**, 629-643.

DAMUTH, J.E., KOLLA, V., FLOOD, R.D., KOWSMANN, R.O., MONTEIRO, M.C., GORINO, M.A., PALMA, J.J.C. & BELDERSON, R.H. 1983a. Distributary channel meandering and bifurcation patterns on the Amazon deep-sea fan as revealed by long-range side-scan sonar (GLORIA). *Geology*, **11**, 94-98.

DAMUTH, J.E., KOWSMANN, R.O., FLOOD, R.D., BELDERSON, R.H. & GORINO, M.A. 1983b. Age relationships of distributary channels on Amazon Deep-Sea Fan: implications for fan growth pattern. Geology, **11**, 470-473.

DAMUTH, J.E. & FLOOD, R.D. 1984. Morphology, sedimentation processes and growth pattern of the Amazon deep-sea fan. *Geo-Marine Letters*, **3**, 109-117.

DAMUTH, J.E. & FLOOD, R.D. 1985. Amazon Fan, Atlantic Ocean. *In:* BOUMA, A.H. BARNES, N.E. & NORMARK, W.R. (eds), *Submarine fans and related turbidite systems*, 97-106. New York: Springer-Verlag.

DAMUTH, J.C., FLOOD, R.D., KOWSMANN, R.O., BELDERSON, R.H. & GORINI, M.A. 1988. Anatomy and growth pattern of Amazon deep-sea fan as revealed by log-range side-scan sonar (GLORIA) and high-resolution seismic studies. *American Association of Petroleum Geologists Bulletin*, **72**, 885-911.

DAMUTH, J.E., FLOOD, R.D., PIRMEZ, C. & MANLEY, P.L. 1995 Architectural elements and depositional processes of Amazon Deep-sea Fan imaged by sidescan sonar (GLORIA), bathymetric swath-mapping (SeaBeam), high-resolution seismic, and piston-core data. *In:* PICKERING, K.T., HISCOTT, R.N., KENYON, N.H., RICCI LUCCHI, F. & SMITH, R.D.A. (eds), *Atlas of Deep Water Environments: Architectural Style in Turbidite Systems*, 105-121. London: Chapman & Hall.

DEEP SEA DRILLING PROJECT LEG 96 SHIPBOARD STAFF 1983. Deep Sea Drilling Project. On the Mississippi Fan. *Nature*, **306**, 736-737.

DROZ, L. 1983. *L'eventail sous-marine profond du Rhone (Gulf du Lion): grands traits morphologiques, et structure semi-profonde.* Ph.D. thesis, Université Pierre & Marie Curie, Paris.

DROZ, L. & BELLAICHE, G. 1985. Rhone deep-sea fan: morphostructure and growth pattern. *American Association of Petroleum Geologists Bulletin*, **69**, 460-479.

DUPUY, G.L.P., OSWALDT, G. & SENS, J. 1963. Champ de Cazaux, géologie et production. *Sixth World Petroleum Congress, Proceedings*. Section 2, 199-213.

EEZ-SCAN 84 SCIENTIFIC STAFF 1986. *Atlas of the US Exclusive Economic Zone, Western conterminous United States.* US Geological Survey Miscellaneous Investigations Series I-1792, scale 1:500,000.

EEZ-SCAN 84 SCIENTIFIC STAFF 1988. Physiography of the western United States Exclusive Economic Zone. *Geology*, **16**, 131-134.

EEZ-SCAN 85 SCIENTIFIC STAFF 1987. *Atlas of the US Exclusive Economic Zone, Gulf of Mexico and Eastern Caribbean Areas.* US Geological Survey Miscellaneous Investigations Series I-1864, 104 scale 1:500,000.

EEZ-SCAN 87 SCIENTIFIC STAFF 1990. *Atlas of the US Exclusive Economic Zone, Atlantic continental margin.* US Geological Survey Miscellaneous Investigations Series I-2054, scale 1:500,000.

EMBLEY, R.W. 1985. A locally formed deep ocean canyon system along the Blanco Transform, northeast Pacific. *GeoMarine Letters*, **5**, 99-104.

EMBLEY, R.W., EWING, J.I., EWING, M. 1970. The Vidal deep-sea channel and its relationship to the Demerara and Barracuda Abyssal Plains. *Deep-Sea Research*, **17**, 539-552.

EINSELE, G. 1985. Responses of sediments to sea-level changes in differing subsiding storm-dominated marginal and epeiric basins. *In:* BAYER, U. & SELIACHER, A. (eds), *Sedimentary and Evolutionary Cycles*, 68-112. Belin: Springer-Verlag.

EINSELE, G., RICKEN, W. & SEILACHER, A. (EDS) 1991. *Cycles and Events in Stratigraphy.* Berlin: Springer-Verlag.

ENGLAND, T.D.J. & HISCOTT, R.N. 1992. Lithostratigraphy and deep-water setting of the upper Nanaimo Group (Upper Cretaceous), outer Gulf Islands of southwestern British Columbia. *Canadian Journal of Earth Science*, **29**, 574-595.

ESTRADA ALIBERAS, M.R. 1982. *Lobulos deposicionales de la parte superior del Grupo de Hecho entre el anticlinal de Boltana y el Rio Aragon (Huesca).* Ph.D. Thesis, Universidad Autonoma de Barcelona, Spain.

FEELEY, M.H. 1984. *Seismic stratigraphic analysis of the Mississippi Fan.* Ph.D. Thesis, Texas A&M University, College Station, Texas.

FIELDING, C.R. & CRANE, R.C. 1987. An application of statistical modelling to the prediction of hydrocarbon recovery factors in fluvial reservoir sequences. *In:* ETHRIDGE, F.G., FLORES, R.M. & HARVEY, M.D. (eds), *Recent developments in fluvial sedimentology*, 321-328. Society of Economic Paleontologists and Mineralogists,

Special Publication, **39**. Tulsa, Oklahoma: Society of Economic Paleontologists and Mineralogists.

FISK, H.N. 1947. *Fine grained alluvial deposits and their effect on Mississippi River activity.* Mississippi River Commission, Vicksberg, Mississippi.

FLEWELLEN, C., MILLARD, N. & ROUSE, I. 1993. TOBI, a vehicle for deep ocean survey. *Electronics and Communication Engineering Journal*, 85-93.

FLOOD, R.D. & DAMUTH, J.E. 1987. Quantitative characteristics of sinuous distributary channels on the Amazon deep-sea fan. *Bulletin of the Geological Society of America*, **98**, 728-738.

FLOOD, R.D., MANLEY, P.C., KOWSMAN, R.O., APPI, C.J. & PIRMEZ, C. 1991. Seismic facies and Late Quaternary growth of Amazon Submarine Fan. *In*: WEIMER, P. & LINK, M.H. (eds), *Seismic facies and sedimentary processes of submarine fans and turbidite systems*, 415-433. New York: Springer-Verlag.

FLOOD, R.D., PIPER, D.J.W., KLAUS, A., ET AL., 1995. *Proceedings of the Ocean Drilling Program, Initial Reports*, **155**. Washington, D.C.: US Government Printing Office.

FRIEND, P.F. 1983. Towards the field classification of alluvial architecture or sequence. *In*: COLLINSON, J.D. & LEWIN, J. (eds), *Modern and Ancient Fluvial Systems*, 345-354. International Association of Sedimentologists Special Publication, **6**.

FRIEND, P.F., SLATER, M.J. & WILLIAMS, R.C., 1979. Vertical and lateral building of river sandstone bodies, Ebro Basin, Spain. *Journal of the Geology Society ,London*, **136**, 39-46.

GALLOWAY, W.E. 1989a. Genetic stratigraphic sequences in basin analysis I: architecture and genesis of flooding-surface bounded depositional units. *American Association of Petroleum Geologists Bulletin*, **73**, 125-142.

GALLOWAY, W.E. 1989b. Genetic stratigraphic sequences in basin analysis II: application to northwest Gulf of Mexico Cenozoic basin. *American Association of Petroleum Geologists Bulletin*, **73**, 143-154.

GALLOWAY, W.E., GARBER, J.L., XIJIN LIU & SLOAN, B.J. 1993. Sequence stratigraphic and depositional framework of the Cenozoic fill, Central and Northern North Sea Basin. *In*: PARKER, J.R. (ed.), *Petroleum Geology of Northwest Europe: Proceedings of the 4th Conference*, 33-43. Bath: The Geological Society.

GARDNER, J.V., FIELD, M.E., LEE, H., EDWARDS, B.E., MASSON, D.G., KENYON, N.H. & KIDD, R.B. 1991. Ground-truthing 6.5kHz side scan sonographs : What are we really imaging? *Journal of Geophysical Research*, **96**, 5955-5974.

GARDNER, M.H. & SONNENFELD, M.D. 1996. Recognition criteria for establishing a high-resolution sequence stratigraphic framework for high net-to-gross slope sandstones Permian Brushy Canyon Formation, TX. *Abstract American Association of Petroleum Geologists National Convention (San Diego)*.

GARRISON, L.E., KENYON, N.H. & BOUMA, A.H. 1982. Channel systems and lobe construction in the Mississippi Fan. *Geo-Marine Letters*, **2**, 31-39.

GHIBAUDO, G. 1992. Subaqueous sediment gravity flow deposits: practical criteria for their field description and classification. *Sedimentology*, **39**, 423-454.

GRIGGS, G.B. & KULM, L.D. 1970. Sedimentation in the Cascadia Deep-Sea Channel. *Bulletin of the Geological Society of America*, **81**, 1361-1384.

GRIGGS, G.B. & KULM, L.D. 1973. Origin and development of Cascadia Deep-Sea Channel. *Journal of Geophyisical Research*, **78**, 6325-6339.

HAMILTON, E.L. 1967. Marine geology abyssal plains in the Gulf of Alaska. *Journal of Geophyisical Research*, **72**, 4189-4213.

HAMPTON, M.A., KARL, H.A. & KENYON N.H. 1989. Sea-floor drainage features of the Cascadia Basin and the adjacent continental slope, Northeast Pacific Ocean. *Marine Geology*, **87**, 249-272.

HAQ, B.U., HARDENBOL, J. & VAIL, P.R. 1987. Chronology of fluctuating sealevels since the Triassic. *Science*, **235**, 1156-1167.

HAQ, B.U., HARDENBOL, J. & VAIL, P.R. 1988. Mesozoic and Cenozoic chronostratigraphy and eustatic cycles. *In*: WILGUS, C.K. ET AL. (eds), *Sea-level research: An integrated approach*, 71-108. Society of Economic Paleontologists and Mineralogists, Special Publication, **42**. Tulsa. Oklahoma: Society of Economic Paleontologists and Mineralogists.

HARTOG JAGER, D., GILES, M.R. & GRIFFITHS, G.R. 1993. Evolution of Paleogene submarine fans of the North Sea in space and time. *In*: PARKER, J.R. (ed.), *Petroleum Geology of Northwest Europe: Proceedings of the 4th Conference*, 59-71. Bath: The Geological Society.

HEEZEN, B.C., THARP, M. & EWING, M. 1959. *The floors of the ocean.* Geological Society of America Special Paper, **65**.

HEEZEN, B.C., JOHNSON, G.L. & HOLLISTER, C.D. 1969. The northwest Atlantic Mid-Ocean Canyon. Journal of Earth Science, **6**, 1441-1453.

HEIN, F.J. & WALKER, R.G. 1982. The Cambro-Ordovician Cap Enragé Formation, Québec, Canada: conglomeratic deposits of a braided submarine channel with terraces. *Sedimentology*, **29**, 309-329.

HELLER, P.L. & DICKENSON, W.R. 1985. Submarine ramp facies model for delta-fed, sand-rich turbidite systems. *American Association of Petroleum Geologists Bulletin*, **69**, 960-976.

HILTON, V. & PICKERING, K.T. 1995. The Montagne de Chalufy turbidite onlap, Eocene-Oligocene turbidite sheet system, Haute Provence, French Maritime Alps. *In*: PICKERING, K.T., HISCOTT, R.N., KENYON, N.H., RICCI LUCCHI, F. & SMITH, R.D.A. (eds), *Atlas of Deep Water Environments: Architectural Style in Turbidite Systems*, 236-241. London: Chapman & Hall.

HILTON, V. & PICKERING, K.T. in prep. Deep-marine high-continuity sand-prone turbidite system, Paleogene Provencal Basin, ystem, Haute Provence, SE. France.

HISCOTT, R.N. 1979. Clastic sills and dikes associated with deep-water sandstones, Tourelle Formation, Ordovician, Quebec. *Journal of Sedimentary Petrology*, **49**, 1-10.

HUGHES CLARKE, J.E. 1988. *The geological record of the 1929 Grand Banks earthquake and its relevance to deep-sea clastic sedimentation*. Ph.D. Thesis, Dalhousie University, Canada.

HUGHES CLARKE, J.E., SHOR, A.N., PIPER, D.J.W. & MAYER, L.A. 1990. Large-scale current-induced erosion and deposition in the path of the 1929 Grand Banks turbidity current. *Sedimentology*, **37**, 613-629.

JACKA, A.D., BECK, R.H., ST GERMAIN, L.C. & HARRISON, S.G. 1968. Permian deep-sea fans of the Delaware Mountain Group (Gaudalupian), Delaware Basin. *In*: SILVER, B.A. (ed.), *Guadalupian facies, Apache Mountain area, west Texas*, 49-90. Permian Basin Section Society of Economic Paleontologists and Mineralogists Publication.

JENSSEN, A.I., BERGSLIEN, D., RYE-LARSEN, M. & LINDHOLM, R.M. 1993. Origin of complex mound geometry of Paleocene submarine-fan reservoirs, Balder Field, Norway. *In*: PARKER, J.R. (ed.), *Petroleum Geology of Northwest Europe: Proceedings of the 4th Conference*, 135-143. Bath: The Geological Society.

KARL, H. A., HAMPTON, M. A. & KENYON, N. H. 1989. Lateral migration of Cascadia deep-sea channel in response to accretionary tectonics. *Geology*, **17**, 144-147.

KASTENS, K.A. & SHOR, A.N. 1985. Depositional processes of a meandering channel on Mississippi Fan. *American Association of Petroleum Geologists Bulletin*, **69**, 190-202.

KENYON, N.H. 1992. Channelised deep-sea siliciclastic systems: a plan view perspective. *Sequence Stratigraphy of European Basins, Dijon. May 18-20 1992*. CNRS/Institute francais du Petroleum Dijon, 458-459.

KENYON, N.H. & MILLINGTON, J. 1995. Contrasting deep-sea depositional systems in the Bering Sea. *In*: PICKERING, K.T., HISCOTT, R.N., KENYON, N.H., RICCI LUCCHI, F. & SMITH, R.D.A. (eds), *Atlas of Deep Water Environments: Architectural Style in Turbidite Systems*, 196-202. London: Chapman & Hall.

KENYON, N.H., AMIR, A. & CRAMP, A. 1995a. Geometry of the younger sediment bodies on the Indus Fan. *In*: PICKERING, K.T., HISCOTT, R.N., KENYON, N.H., RICCI LUCCHI, F. & SMITH, R.D.A. (eds), *Atlas of Deep Water Environments: Architectural Style in Turbidite Systems*, 89-93 London: Chapman & Hall.

KENYON, N.H., MILLINGTON, J., DROZ, L. & IVANOV, M.K. 1995b. Scour holes in a channel-lobe transition zone on the Rhone Cone. *In*: PICKERING, K.T., HISCOTT, R.N., KENYON, N.H., RICCI LUCCHI, F. & SMITH, R.D.A. (eds), *Atlas of Deep Water Environments: Architectural Style in Turbidite Systems*, 212-215. London: Chapman & Hall.

KLEVERLAAN, K. 1987. Gordo Megabed: a possible seismite in a Tortonian submarine fan, Tabernas Basin, Province Almeria, SE Spain. *Sedimentary Geology*, **51**, 165-180.

KLEVERLAAN, K. 1989. Three distinctive feeder-lobe systems within one time slice of the Tortonian Tabernas fan, SE Spain. *Sedimentology*, **36**, 25-45.

KOLLA, V. 1993. Lowstand deep-water siliciclastic depositional systems: characteristics and terminologies in sequence stratigraphy and sedimentology. *Bulletin Centres Recherches Exploration-Production Elf Aquitaine*, **17**, 67-78.

KOLLA, V. & COUMES, F. 1985. Morpho-acoustic and sedimentologic characteristics of the Indus Fan. *Geo-Marine Letters*, **3**, 133-139.

KOLLA, V. & F. COUMES 1987. Morphology, internal structure, seismic stratigraphy, and sedimentation of Indus Fan. *American Association Petroleum Geologists Bulletin*, **71**, 650-77.

KOLLA, V. & MACURDA JR., D.B. 1988. Sea-level changes and timing of turbidity-current events in deep-sea fan systems. *In*: WILGUS, C.K., HASTINGS, B.S., KENDAL, C.G.C., POSAMENTIER, H.W., ROSS, C.A. & VAN WAGONER, J.C. (eds), *Sea level changes: an integrated approach*, 381-392. Society of Economic Paleontologists and Mineralogists Special Publication, **42**. Tulsa, Oklahoma: Society of Economic Paleontologists and Mineralogists.

KOLLA, V. & PERLMUTTER, M.A. 1993. Timing of turbidite sedimentation on the Mississippi Fan. *American Association of Petroleum Geologists Bulletin*, **77**, 1129-1141.

KOMAR, P.D. 1969. The channelized flow of turbidity currents with application to Monterey deep-sea fan channel. *Journal of Geophysical Research*, **74**, 4544-4558.

KOMAR, P.D. 1973. Continuity of turbidity current flow and systematic variations in deep-sea channel morphology. *Geological Society of America Bulletin*, **84**, 3329-3338.

KULM, L.D., VON HUENE, R.E., ET AL. 1973. *Initial Reports Deep Sea Drilling Project*, **1 8**. Washington, D.C.: US Government Printing Office.

LANGBEIN, W.B. & LEOPOLD, L.B. 1966. River meanders - theory of minimum variance. *U.S. Geological Survey. Professional Paper*, **422-H**.

LAUGHTON, A.S. 1968. New evidence of erosion on the deep sea floor. *Deep-Sea Research*, **15**, 21-30.

LEOPOLD, L.B. & WOLMAN, M.G. 1957. River channel patterns: braided, meandering, and straight. *US Geological survey Professional Paper*, **282-B**, 39-85.

LEOPOLD, L.B. & WOLMAN, M.G. 1960. River meanders. *Bulletin of The Geological Society of America*, **71**, 769-794.

LINK, M.H. & SQUIRES, R.L. 1981. Field Guide. *In*: LINK, M.H., SQUIRES, R.L. & COLBURN, I.P. (eds), *Simi Hills Cretaceous turbidites, southern California*, Pacific Section, Society of Economic Paleontologists and Mineralogists, Fall Field trip Guidebook 111-114.

LINK, M.H., SQUIRES, R.L. & COLBURN, I.P. 1984. Slope and deep-sea fan facies and paleogeography of Upper Cretaceous Chatsworth Formation, Simi Hills, California. *American Association of Petroleum Geologist Bulletin*, **68**, 850-873.

LISITZIN, A.P. 1972. *Sedimentation in the world ocean*. Society of Economic Paleontologists and Mineralogists Special Publication, **17**. Tulsa, Oklahoma: Society of Economic Paleontologists and Mineralogists.

LOWE, D.R. 1982. Sediment gravity flows: II. Depositional models with special reference to the deposits of high-density turbidity currents. *Journal of Sedimentary Petrology*, **52**, 279-297.

MALINVERNO, A., RYAN, W.B.F., AUFFRET, G. & PAUTOT, G. 1988. Sonar images of the path of recent failure events on the continental margin off Nice, France. *Geological Society of America Special Paper*, **229**, 59-75.

MANLEY, P.L. & FLOOD, R.D. 1988. Cyclic sediment deposition within Amazon deep-sea fan. *American Association of Petroleum Geologists Bulletin*, **72**, 912-925.

MASSON, D.G., GARDNER, J.V., PARSON, L.M. & FIELD, M.E. 1985. Morphology of upper Laurentian Fan using GLORIA long-range sidescan sonar. *American Association of Petroleum Geologists Bulletin*, **69**, 950-959.

MASSON, D.G., HUGGETT, Q.J., WEAVER, P.P.E., KIDD, R.B. & GARDNER, J.V. 1991. Quaternary sedimentary processes on the N.W. African Continental Margin - an integrated study using sidescan sonar, high-resolution profiling and core data. *American Association of Petroleum Geologists Bulletin*, **75**, 1416.

MASSON, D.G., KENYON N.H., GARDNER, J.V. & FIELD, M.E. 1995. Monterey Fan: channel and overbank morphology. *In*: PICKERING, K.T., HISCOTT, R.N., KENYON, N.H., RICCI LUCCHI, F. & SMITH, R.D.A. (eds), *Atlas of Deep Water Environments: Architectural Style in Turbidite Systems*, 74-79. London: Chapman & Hall.

MCGREGOR, B.A., STUBBLEFIELD, W.L., RYAN, W.B.F. & TWICHELL, D.C. 1982. Wilmington Submarine canyon: a marine fluvial-like system. *Geology*, **10**, 27-30.

MCHARGUE, T.R. 1991. Sesmic facies, processes and evolution of Miocene inner fan channels, Indus Submarine Fan. *In*: WEIMER, P. & LINK, M.H. (eds), *Seismic facies and sedimentary processes of submarine fans and turbidite systems*, 403-414. New York: Springer-Verlag.

MCHARGUE, T.R. & WEBB, J.E. 1986. Internal geometry, seismic facies and petroleum potential of canyons and inner fan channels of the Indus Submarine Fan. *American Association of Petroleum Geologists Bulletin*, **70**, 161-180.

MCKEE, E.D. & WEIR, G.W., 1953. Terminology of stratification and cross-stratification. *Geology Society of America Bulletin*, **64**, 381-390.

MCKEE, E.D., CROSBY, E.J. & BERRYHILL, H.L. JR. 1967. Flood deposits, Bijou Creek, Colorado, June 1965. *Journal of sedimentary Petrology*, **37**, 829-851.

MENARD, H.W. 1955. Deep-sea channels, topography, and sedimentation. *American Association of Petroleum Geologists Bulletin*, **39**, 236-255.

MENARD, H.W. 1964. *Marine Geology of the Pacific*. New York: McGraw Hill.

MIALL, A.D. 1985. Architectural-element analysis: a new method of facies analysis applied to fluvial deposits, *Earth-Science Reviews* **22**, 261-308.

MIALL, A.D. 1989. Architectural elements and bounding surfaces in channelized clastic deposits: notes on comparisons between fluvial and turbidite systems. *In*: TAIRA, A. & MASUDA, F. (eds), *Sedimentary Facies in the Active Plate Margi*, 3-15. Tokyo: Terra Scientific Publishing Company (TERRAPUB).

MIALL, A.D. 1992. Exxon global cycle chart: An event for every occasion? *Geology*, **20**, 787-790.

MIDDLETON, G.V. 1966a. Experiments on density and turbidity currents: I. Motion of the head. *Canadian Journal of Earth Sciences*, **3**, 523-546.

MIDDLETON, G.V. 1966b. Experiments on density and turbidity currents: II. Uniform flow of density currents. *Canadian Journal of Earth Sciences*, **3**, 627-637.

MIDDLETON, G.V. 1966c. Small scale models of turbidity currents and the criterion for auto-suspension. *Journal of Sedimentary Petrology*, **36**, 202-208.

MIDDLETON, G.V. & SOUTHARD, J.B. 1984. *Mechanics of sediment transport*. Revised Short Course Notes. Tusla, Oklahoma: Society of Economic Paleontologists and Mineralogists.

MILLINGTON, J. & CLARK, J.D. 1995. Submarine canyon and associated base-of-slope sheet system: the Eocene Charo-Arro system south-central Pyrenees. Spain. *In*: PICKERING, K.T., HISCOTT, R.N., KENYON, N.H., RICCI LUCCHI, F. & SMITH, R.D.A. (eds), *Atlas of Deep Water Environments: Architectural Style in Turbidite Systems*, 150-156. London: Chapman & Hall.

MILLINGTON, J. & CLARK, J.D. 1996. The Charo/Arro canyon-mouth sheet system, south-central Pyrenees, Spain: a structurally influenced zone of sediment dispersal. *Journal of Sedimentary Research*, **B65**, 443-454.

MITCHUM R.M. 1977. Seismic stratigraphy and global changes of sea level, Part 1: Glossary of terms used in seismic stratigraphy. *In:* PAYTON, C.E. (ed.), *Seismic stratigraphy - applications to hydrocarbon exploration*, 205-212. American Association of of Petroleum Geologists Memoir, **26**. Tulsa, Oklahoma: Society of Economic Paleontologists and Mineralogists.

MITCHUM, R.M., VAIL, P.R. & THOMPSON III, S. 1977. Seismic stratigraphy and global changes of sea level, Part 2: the depositional sequence as a basic unit for stratigraphic analysis. *In:* PAYTON, C.E. (ed.), *Seismic stratigraphy - applications to hydrocarbon exploration*, American Association of Petroleum Geologists Memoir, Tulsa, **26**, 53-62.

MITCHUM, R.M. JR. 1985. Seismic stratigraphic expression of submarine fans. *In*: BERG O.R. & WOOLVERTON D.G. (eds), *Seismic stratigraphy II: an integrated approach to hydrocarbon exploration*, 117-136. American Association of Petroleum Geologists Memoir, **39**.

MOORE, G.T., STARKE, G.W., BONHAM, L.C. ET AL. 1978. Mississippi Fan, Gulf of Mexico - physiography, stratigraphy and sedimentation patterns. *In*: BOUMA, A.H., MOORE, G.T. & COLEMAN, J.M. (eds), *Framework, facies and oil-trapping characteristics on upper continental margin*, 155-199. American Association of Petroleum Geologists Studies in Geology, **7**.

MUTTI, E. 1977. Distinctive thin-bedded turbidite facies and related depositional environments in the Eocene Hecho Group (south-central Pyrenees, Spain). *Sedimentology*, **24**, 107-31.

MUTTI, E. 1979. Turbidites et cones sous-marins profunds. *In*: HOMEWOOD, P. (ed.), *Sedimentation detritique (fluviatile, littorale et marine)*. 353-419. Fribourg, Switzerland: Institut Geologique University.

MUTTI, E. 1992. *Turbidite Sandstones*. Agip, Italy.

MUTTI, E. & RICCI LUCCHI, F. 1972. Le torbidit dell'Apennino settentrionale: introduzione all'analisi di facies. *Memoir Society of Geology Italy*, **11**, 161-199. (English translation by NILSEN, T.H. 1978. *International Geological Review*, **20**, 125-166).

MUTTI, E. LUTERBACHER, H.P., FERRER, J. & ROSELL, J. 1972. Schema stratigrafica e lineamenti di facies del Paleogene marino della Zona Centrale Sud-Prenaica tra Tremp (Catalogna) e Pamplona (Navarra). *Memoir of the Geological Society of Italy*, **11**, 391-416.

MUTTI, E. & RICCI LUCCHI, F. 1975. Turbidite facies and facies associations. *In*: MUTTI, E., PAREA, G.C., RICCI-LUCCHI, F., SAGRI, M., ZANAUCCHI, G., CHIBAUDOAND, G. & IRACCARION, S. (eds), *Examples of turbidite facies and facies associations from selected formations of the Northern Apennines*, 21-36. IX International Congress on Sedimentologists, Nice, Field Trip A-11.

MUTTI, E. & M. SONNINO 1981. Compensation cycles: a diagnostic feature of turbidite sandstone lobes. *In: Abstracts Volume, 120-123. 2nd European Regional Meeting, Bologna, Italy.* International Association of Sedimentologists.

MUTTI, E., REMACHA, E., SGAVETTI, M., ROSELL, J., VALLONI, R. & ZAMORANO, M. 1985. Stratigraphy and facies characteristics of the Eocene Hecho Group turbidite systems, South Central Pyrenees. *In:* MILA, M.D. & ROSELL, J. (eds), *Excursion guidebook of the 6th European regional meeting of the International Association of Sedimentologists,* Llerida, 521-576. International Association of Sedimentologists.

MUTTI, E. & NORMARK, W.R. 1987. Comparing examples of modern and ancient turbidite systems: problems and concepts. *In:* LEGGET, J.K. & ZUFFA, G.G. (eds), *Marine Clastic Sedimentology: Concepts and Case Studies,* 1-38. London: Graham & Trotman.

MUTTI, E., SEGURET, M. & SGAVETTI, M. 1989. Sedimentation and deformation in the Tertiary Sequences of the Southern Pyrenees. *American Association of Petroleum Geologists Mediterranean Basins Conference Guidebook Field Trip No. 7, Nice.* Special Publication of the Institute of Geology, University of Parma.

MUTTI, E. & NORMARK, W.R. 1991. An integrated approach to the study of turbidite systems. *In:* WEIMER, P. & LINK, M.H. (eds), *Seismic facies and sedimentary processes of submarine fans and turbidite systems,* 75-106. New York: Springer-Verlag.

NELSON, C.H., 1985. Astoria Fan, Pacific Ocean. *In:* BOUMA, A.H. BARNES, N.E. & NORMARK, W.R. (eds),, *Submarine fans and related turbidite systems,* 45-50. New York: Springer-Verlag.

NELSON, C.H., CARLSON, P.R., BYRNE, J.V. & ALPHA, T.R. 1970. Development of the Astoria Canyon - fan physiography and comparison with similar systems. *Marine Geology,* **8,** 259-291.

NELSON, C.N. & KULM, L.D. 1973. Submarine fans and deep-sea channels. *In:* MIDDLETON, G.V. & BOUMA, A.H. (eds), *Turbidites and Deep Water Sedimentation,* 39-78. Society of Economic Paleontolologists and Mineralogists Pacific Section Short Course, Anaheim. Tulsa, Oklahoma: Society of Economic Paleontologists and Mineralogists.

NELSON, C.H. & NILSEN, T. 1974. Depositional trends of modern and ancient deep-sea fans. *In:* DOTT, R.H. & SHAVER, R.H. (eds), *Modern and ancient geosynclinal sedimentation,* 69-91. Special Publication of the Society of Economic Paleontologists and Mineralogists, **19.** Tulsa, Oklahoma: Society of Economic Paleontologists and Mineralogists.

NELSON, C.H. & NILSEN, T.H. 1984. Modern and ancient deep-sea fan sedimentation. Society of Economic Paleontolologists and Mineralogists Short Course, **14.**

NEWMAN, M. ST. J., REEDER, M.L., WOODRUFF, A.H.W. & HATTON, I.R. 1993. The geology of the Gryphon Field. *In:* PARKER, J.E. (ed.) *In:* PARKER, J.R. (ed.), *Petroleum Geology of Northwest Europe: Proceedings of the 4th Conference,* 123-133. Bath: The Geological Society.

NEWTON, S.K. & FLANAGAN, K.P. 1993. The Alba Field: evolution of the depositional model. *In:* PARKER, J.E. (ed.) *In:* PARKER, J.R. (ed.), *Petroleum Geology of Northwest Europe: Proceedings of the 4th Conference,* 161-171. Bath: The Geological Society.

NESS, G.E. 1972. *The structure and sediments of Surveyor Deep-Sea Channel.* M.S. thesis, Oregon State University, Corvallis, Oregon.

NESS, G.E. & KULM, L.D. 1973. Origin and development of Surveyor Deep-Sea Channel. Bulletin of the Geological Society of America, **84,** 3339-3354.

NILSEN, T.H. 1985. Chugach turbidite system, Alaska. *In:* BOUMA, A.H. BARNES, N.E. & NORMARK, W.R. (eds), *Submarine fans and related turbidite systems,* 185-192. New York: Springer-Verlag.

NILSEN, T.H. & ABBOTT, P.L. 1981. Paleogeography and sedimentology of Upper Cretaceous turbidites, San Diego, California. *American Association of Petroleum Geologists Bulletin,* **65,** 1256-1284.

NORMARK, W.R. 1970a. Growth patterns of deep-sea fans. *American Association of Petroleum Geologists Bulletin,* **54,** 2170-2195.

NORMARK, W.R. 1970b. Channel piracy on Monterey Deep-Sea Fan. *Deep Sea Research,* **17,** 837-846.

NORMARK, W.R. 1978. Fan valleys, channels, and depositional lobes on modern submarine fans: characters for recognition of sandy turbidite environments. *American Association of Petroleum Geologists Bulletin,* **62,** 912-931.

NORMARK, W.R. 1985. Local morphologic controls and effects of basin geometry on flow processes in deep marine basins. *In:* ZUFFA, G.G. (ed.), *Provenance of Arenites,* 47-63. Dordrecht, Amsterdam: D.Riedel.

NORMARK, W.R. & PIPER, D.J.W. 1969. Deep-sea fan-valleys, past and present. *Bulletin of the Geological Society of America*, **80**, 1859-1866.

NORMARK, W.R., PIPER, D.J.W. & HESSE G.R. 1979. Distributary channels, sand lobes and mesotopography of Navy submarine fan, California Borderland, with application to ancient fan sediments. *Sedimentology*, **26**, 749-774.

NORMARK, W.R., HESS, G.R., STOW, D.A.V. & BOWEN, A.J. 1980. Sediment waves on the Monterey Fan levee: a preliminary physical interpretation. *Marine Geology*, **37**, 1-18.

NORMARK, W.R., BARNES, N.E. & COUMES, F. 1984. Rhone deep sea fan: a review. *Geo-Marine Letters*, **3**, 155-160.

NORMARK, W.R. & GUTMACHER, C.E. 1985. Delgada Fan, Pacific Ocean. *In:* BOUMA, A.H. BARNES, N.E. & NORMARK, W.R. (eds), *Submarine fans and related turbidite systems*, 59-64. New York: Springer-Verlag.

NORMARK, W.R. & PIPER, D.J.W. 1991. Initiation processes and flow evolution of turbidity currents: implications for the depositional record. *In:* OSBORNE, R.H. (ed.), *From shoreline to abyss: contributions in marine geology in honor of Francis Parker Shepard*, 207-230. Society of Economic Paleontologists and Mineralogists Special Publication, **46**. Tulsa, Oklahoma: Society of Economic Paleontologists and Mineralogists.

O'CONNELL, S., STELTING, C.E., BOUMA, A.H., COLMAN, J. M., CREMER, M., DROZ, L., MEYER-WRIGHT, A.A., NORMARK, W. R., PICKERING, K. T., STOW, D. A. V. & DSDP LEG 96 SHIPBOARD SCIENTISTS 1985. Drilling results on the Lower Mississippi Fan. *In:* BOUMA, A.H. BARNES, N.E. & NORMARK, W.R. (eds), *Submarine fans and related turbidite systems*, 291-298. New York: Springer-Verlag.

O'CONNELL, S. & NORMARK, W.R. 1986. Acoustic facies and sediment composition of the Mississippi Fan drill sites. *In:* BOUMA, A.H., COLEMAN, J.M., ET AL. (eds), *Initial reports of the Deep Sea Drilling Project*, **96**, 457-473. Washington, D.C.: US Government Printing Office.

O'CONNELL, S., NORMARK, W.R., RYAN, W.B.F. & KENYON, N.H. 1991. An entrenched thalweg channel on the Rhône Fan: interpretation from a SEABEAM and SEAMARC I survey. *In:* OSBORNE, R.H. (ed.), *From shoreline to abyss: contributions in marine geology in honor of Francis Parker Shepard*, 259-270. Society of Economic Paleontologists and Mineralogists Special Publication, **46**. Tulsa, Oklahoma: Society of Economic Paleontologists and Mineralogists.

ORI, G.G. & FRIEND, P.F. 1984. Sedimentary basins formed and carried piggy-back on active thrust sheets: *Geology,* **12**, 475-478.

PANTIN, H.M. 1979. Interaction between velocity and effective density in turbidity flow: phase plane analysis, with criteria for autosuspension. *Marine Geology,* **31**, 59-99.

PHILIPS, S. 1987. Dipmeter interpretation of turbidite-channel reservoir sandstones, Indian Draw Field, New Mexico. *In:* TILLMAN, R.W. & WEBER, K.J. (eds), *Reservoir Sedimentology*, 113-128. Society of Economic Paleontologists and Mineralogists Special Publication **40**. Tulsa, Oklahoma: Society of Economic Paleontologists and Mineralogists.

PICKERING, K.T. 1979a. *A Precambrian submarine fan and upper basin slope succession in the Barents Sea Group, Finnmark, North Norway.* Unpublished D.Phil. Thesis, University of Oxford, UK.

PICKERING, K.T. 1979b. Possible retrogressive flow slide deposits from the Kongsfjord Formation: a Precambrian submarine fan, Finnmark, N. Norway. *Sedimentology,* **26**, 295-306.

PICKERING, K.T. 1981. Two types of outer fan lobe sequence, from the late Precambrian Kongsfjord Formation submarine fan, Finnmark, north Norway. *Journal of Sedimentary Petrology,* **51**, 1277-1286.

PICKERING, K.T. 1982a. Middle-fan deposits from the late Precambrian Kongsfjord Formation submarine fan, northeast Finnmark, northern Norway. *Sedimentary Geology,* **33**, 79-110.

PICKERING, K.T. 1982b. The shape of deep-water siliciclastic systems: a discussion. *Geomarine Letters,* **2**, 41-47.

PICKERING, K.T. 1983. Transitional submarine fan deposits from the late Precambrian Kongsfjord Formation submarine fan, NE Finnmark, N. Norway. *Sedimentology,* **30**, 181-199.

PICKERING, K.T. 1985. Kongsfjord turbidite system, Norway. *In:* BOUMA, A.H. BARNES, N.E. & NORMARK, W.R. (eds), *Submarine fans and related turbidite systems*, 237-244. New York: Springer-Verlag.

PICKERING, K.T., STOW, D.A.V., WATSON, M.P. & HISCOTT, R.N. 1986a. Deep-water facies, processes and models: a review and classification scheme for modern and ancient sediments. *Earth Science Reviews* , **23**, 75-174.

PICKERING, K.T., COLMAN, J. M., CREMER, M., DROZ, L., KOHL, B., NORMARK, W.R., O'CONNELL, S., STOW, D.A.V. & MEYER-WRIGHT, A. 1986b. A high sinuosity, laterally migrating submarine fan channel-levee-overbank: results from DSDP Leg 96 on the Mississippi Fan, Gulf of Mexico. *Marine and Petroleum Geology*, **3**, 3-18.

PICKERING, K.T., HISCOTT, R.N. & HEIN, F.J. 1989. *Deep marine environments: clastic sedimentation and tectonics.* London: Chapman & Hall.

PICKERING, K.T., HISCOTT, R.N., KENYON, N.H., RICCI LUCCHI, F. & SMITH, R.D.A. (eds), 1995a. *Atlas of Deep Water Environments: Architectural Style in Turbidite Systems.* London: Chapman & Hall.

PICKERING, K.T., CLARK, J.D., SMITH, R.D.A, HISCOTT, R.N., RICCI LUCCHI, F. & KENYON, N.H. 1995b. Architectural element analysis of turbidite systems, and selected topical problems for sand-prone deep-water syatems. *In*: PICKERING, K.T., HISCOTT, R.N., KENYON, N.H., RICCI LUCCHI, F. & SMITH, R.D.A. (eds), *Atlas of Deep Water Environments: Architectural Style in Turbidite Systems*, 1-10. London: Chapman & Hall.

PIPER, D.J.W. & NORMARK, W.R. 1983. Turbidite depositional patterns and flow characteristics, Navy submarine fan, California Borderland. *Sedimentology*, **30**, 681-694.

PIPER, D.J.W., STOW, D.A.V. & NORMARK, W.R. 1984. The Laurentian Fan: Sohm Abyssal Plain. *Geo-Marine Letters*, **3**, 141-146.

PIPER, D.J.W., SHOR, A.N., FARRE, J.A., O'CONNELL, S. & JACOBI, R. 1985. Sediment slides and turbidity currents on the Laurentian Fan: side scan sonar investigations near the epicenter of the 1929 Grand Banks earthquake. *Geology*, **13**, 538-541.

PIPER, D.J.W., SHOR, A.N. & HUGHES CLARKE, J.E. 1988. The 1929 Grand Banks earthquake slump and turbidity current. *Geological Society of America Special Paper*, **229**, 77-92.

PRATSON, L.F. & LAINE, E.P. 1989. The relative importance of gravity-induced versus current-controlled sedimentation during the Quaternary along the Mideast U.S. outer continental margin revealed by 3.5 kHz echo character. *Marine Geology*, **89**, 87-126.

PRATSON, L.F., RYAN, W.B.F., MOUNTAIN, G.S. & TWICHELL, D.C. 1994. Submarine canyon initiation by downslope-eroding sediment flows: evidence in Late Cenozoic strata on the New Jersey continental slope. *Geological Society of America Bulletin*, **106**, 395-412.

PRIOR, D.B., ADAMS, C.E. & COLEMAN, J.M. 1983. Characterisation of a deep-sea channel on the Mississippi Fan as revealed by a high-resolution survey. *Gulf Coast Association of Geological Societies Transactions*, **33**, 389-394.

QUINLAN, G.M. & BEAUMONT, C. 1984. Appalation thrusting, lithospheric flexure, and the Palaeozoic stratigraphy of the Eastern Interior of North America. *Canadian Journal of Earth Sciences*, 21, 973-996.

READING, H.G. 1991. The classification of deep-sea depositional systems by sediment calibre and feeder systems. *Journal of the Geological Society, London*, **148**, 427-430.

READING, H.G. & RICHARDS, M. 1994. Turbidite systems in deep-water basin margins classified by grain size and feeder system. *American Association of Petroleum Geologists Bulletin*, **78**, 792-822.

REIMNITZ, E. & GUTIERREZ-ESTRADA, M. 1970. Rapid change in the head of the Rio Balsas submarine canyon system. *Marine Geology*, **8**, 245-258.

REMACHA, G., OMS, O. & COELLO, J. 1995. The Rapitan turbidite channel and its related eastern levee-overbank deposits, Eocene Hecho Group, south-central Pyrenees, Spain. *In*: PICKERING, K.T., HISCOTT, R.N., KENYON, N.H., RICCI LUCCHI, F. & SMITH, R.D.A. (eds), *Atlas of Deep Water Environments: Architectural Style in Turbidite Systems*, 145-149. London: Chapman & Hall.

RENARD, V. & ALLENOU, J.P. 1979. SEABEAM, multibeam echo-sounding in Jean Charcot: description, evaluation and first results. *International Hydrographic Review*, **56**, 35-67.

RICCI LUCCHI, F. 1969. Channelized deposits in the middle Miocene flysch of Romagna (Italy). Giornale di Geologia, **36**, 203-282. (In English).

RICCI LUCCHI, F. 1981. The Marnoso-arenacea: a migrating turbidite basin (over-supplied) by a highly efficient dispersal system. *In*: RICCI LUCCHI, F. (ed.), *Excursion guidebook*, 157-160. 2nd International Association of Sedimentologists European Regional Meeting, Bologna, Abstracts.

SARG, J.F. & LEHMANN, P.J. 1986. Lower and Middle Guadalupian facies and stratigraphy, San Andres/Grayburg formations, Permian Basin, Guadalupe Mountains, New Mexico. *In*: MOORE, G.E. & WILDE, G.L. (eds), *Lower and Middle Guadalupain facies, stratigraphy, and reservoir geometries, San Andres/Grayburg formations, Permian Basin, Guadalupe Mountains, New Mexico and Texas*, 1-5. Permian Basin Section, Society of Economic Paleontologists and Mineralogists Special Publication **86-25**. Tulsa, Oklahoma: Society of Economic Paleontologists and Mineralogists.

SATTERFIELD, W.M. & BEHRENS, E.W. 1990. A late Quaternary canyon/channel system, northwest Gulf of Mexico continental slope. *Marine Geology*, **92**, 51-67.

SCHUMM, S.A. 1963. Sinuosity of alluvial rivers on the Great Plains. *Bulletin of the Geological Society of America*, **74**, 1089-1100.

SCHUMM, S.A. 1981. *Evolution and response of the fluvial system, sedimentologic implications.* Society of Economic Palaeontologists and Mineralogists Special Publication, **31**, 19-29.

SCHUMM, S.A. & KHAN., H.R. 1972. Experimental study of channel patterns. *Bulletin of The Geological Society of America*, **88**, 1755-1770.

SCHUMM, S.A., KHAN., H.R., WINKLEY, B.R. & ROBINS, L.G. 1972. Variability of river patterns: *Nature; Physical Science,* **237**, 75-76.

SGAVETTI, M. 1991. Photostatigraphy of Ancient Turbidite Systems. *In:* WEIMER, P. & LINK, M.H. (eds), *Seismic facies and sedimentary processes of submarine fans and turbidite systems,* 107-126. New York: Springer-Verlag.

SHANMUGAM, G. & MOIOLA, R.J. 1985. Submarine fan models: problems and solutions. *In:* BOUMA, A.H. BARNES, N.E. & NORMARK, W.R. (eds), *Submarine fans and related turbidite systems,* 29-34. New York: Springer-Verlag.

SHANMUGAM, G., J.E. DAMUTH & R.J. MOIOLA 1985. Is the turbidite facies association scheme valid for interpreting ancient submarine fan environments? *Geology* , **13**, 234-237.

SHANMUGAM, G. & MOIOLA, R.J. 1988. Submarine Fans: characteristics, models, classification and reservoir potential. *Earth-Science Reviews*, **24**, 383-428.

SHANMUGAM, G., MOIOLA, R.J., MCPHERSON, J.G. & O'CONNELL, S. 1988. Comparisons of turbiditic facies associations in modern passive-margin Mississippi fan with ancient active-margin fans. *Sedimentary Petrology,* **58**, 63-77.

SHAUB, F.J., BUFFLER, R.T. & PARSONS, J.G. 1984. Seismic stratigraphic framework of deep central Gulf of Mexico Basin. *American Association of Petroleum Geologists Bulletin*, **68**, 1790-1802.

SHEPARD, F.P., PHLEGER, F.B. & VAN ANDEL, T.H. 1960. *Recent sediments, north-west Gulf of Mexico.* American Association of Petroleum Geologists, Tulsa, Oklahoma.

SHOR, A.N., PIPER, D.J.W., HUGHES CLARKE, J. & MEYER, L.A. 1990. Giant flute-like scours and other erosional features formed by the 1929 Grand Banks turbidity current. *Sedimentology*, **37**, 631-645.

SILVER, B.A. & TODD, R.G. 1969. Permian cyclic strata, northern Midland and Delaware basins, west Texas and southeastern New Mexico. *American Association of Petroleum Geologists Bulletin*, **53**, 2223-2251.

SIMPSON, J.E. 1972. Effects of the lower boundary on the head of a gravity current. *Journal of Fluid Mechanics,* **53**, 759-768.

SMITH, R.D.A., WATERS, R. & DAVIES, J. 1991. Upper Ordovician and Lower Silurian turbidite systems in the Welsh Basin. *International Sedimentological Congress, Nottingham, Field Trip Guide*, **20**.

SOMMERS, M.L., CARSON, R.M., REVIE, J.A., EDGE, R.H., BARROW, B.J. & ANDREWS, A.G. 1978. GLORIA II - an improved long-range side-scan sonar. Oceanology International 78, Technical Sessions J, BPS Exhibitions, London, 16-24.

SOMERS, M.L. & SEARLE, R. 1984. GLORIA sounds out the seabed. *New Scientist,* **104**, No. 1428, 12-15.

SONNENFELD, M.D. & GARDNER, M.H. 1996. Slope discordance cycles - key to erecting a stratigraphic hierarchy in slope systems: Middle Permian Brushy Canyon Formation, Guadalupe Mountains, TX. *Abstract AAPG National Convention (San Diego)*

STEFFENS, G. 1986. Pleistocene entrenched valley/canyon systems, Gulf of Mexico. *American Association of Petroleum Geologists Bulletin*, **70**, 1189.

STELTING, C.E. & DSDP LEG 96 SHIPBOARD SCIENTISTS 1985. Migratory characteristics of a mid-fan meander belt, Mississippi Fan. *In:* BOUMA, A.H., NORMARK, W.R. & BARNES, N.E. (eds), *Submarine fans and related turbidite systems,* 283-290. New York: Springer-Verlag.

STELTING, C.E., DROZ, L., BOUMA, A.H., COLEMAN, J.M., CREMER, M., MEYER, A.W., NORMARK, W.R., O'CONNELL, S. & STOW, D.A.V. 1986. Late Pliestocene seismic stratigraphy of the Mississippi Fan. *In:* BOUMA, A.H., COLEMAN, J.M. & MEYER, A.W. ET AL. (eds), *Initial Reports Deep Sea Drilling Project,* **96**, 563-576. Washington, D.C.: U.S. Government Printing Office.

STOCCHI, S. 1987. *Il sistema torbiditico Guaso nella sequenza depositionale di Ainsa (Eocene medio della Zona Centrale Sud-Pirenaica, Provincia di Huesca, Spagna).* Ph.D. Thesis, University of Parma, Italy.

STOW, D.A.V. 1985. Deep-sea clastics: where are we and where are we going? *In:* BRENCHLEY, P.J. &

WILLIAMS, B.P.J. (eds), *Sedimentology: recent developments and applied aspects*, 67-93. Geology Society of London Special Publication, **18**. Oxford: Blackwell Scientific Publications.

STOW, D.A.V. 1986. Deep clastic seas. *In*: READING, H.G. (ed.), *Sedimentary environments and facies 2nd Edition.*, 399-444. Oxford: Blackwell Scientific Publications.

STOW, D.A.V. & A.J. BOWEN 1980. A physical model for the transport and sorting of fine-grained sediments by turbidity currents. *Sedimentology*, **27**, 31-46.

STOW, D.A.V., CREMER, M., DROZ, L., NORMARK, W.R., O'CONNELL, S., PICKERING, K.T., STELTING, C.E., MEYER-WRIGHT, A.A. & DSDP LEG 96 SHIPBOARD SCIENTISTS 1985. Mississippi Fan sedimentary facies, composition and texture. *In*: BOUMA, A.H. BARNES, N.E. & NORMARK, W.R. (eds), *Submarine fans and related turbidite systems*, 259-266. New York: Springer-Verlag.

STUART, C.J. & CAUGHEY, C.A. 1976. Form and composition of the Mississippi fan. *Gulf Coast Association Geological Society Transactions, **26**, 333-343.

SURLYK, F. 1987. Slope and deep shelf gully sandstones, Upper Jurassic, East Greenland. *American Association of Petroleum Geologists Bulletin*, **71**, 464-75.

TAYLOR, G. & WOODYER, K.D. 1978. Bank deposition in suspended load streams. *In*: MIALL, A.D. (ed.), *Fluvial sedimentology*, 257-275. Memoir of the Canadian Society of Petroleum Geologists, Calgary, **5**.

TIMBRELL, G. 1993. Sandstone architecture of the Balder Formation depositional system, UK Quadrant 9 and adjacent areas. *In*: PARKER, J.R. (ed.) *In*: PARKER, J.R. (ed.), *Petroleum Geology of Northwest Europe: Proceedings of the 4th Conference*, 59-71. Bath: The Geological Society.

TWICHELL, D.C., ROBERTS, D.G. & TELEKI, P.G. 1980. Long-range sidescan sonar views of the continental margin seaward of the Baltimore Canyon Trough. *Geological Society of America 93rd Annual Meeting Abstracts with Programs*, **12**, 538-539.

TWICHELL, D.C. & ROBERTS, D.G. 1982. Morphology, distribution, and development of submarine canyons on the United States Atlantic continental slope between Hudson and Baltimore Canyons. *Geology*, **10**, 408-412.

TWICHELL, D.C., KENYON, N.H., PARSON, L.M. & MCGREGOR, B.A. 1991. Depositional patterns of the Mississippi Fan surface: Evidence from GLORIA II and high-resolution seismic profiles.

In: WEIMER, P. & LINK, M.H. (eds),, *Seismic facies and sedimentary processes of submarine fans and turbidite systems*, 349-364. New York: Springer-Verlag

VAN ANDEL, T.J. & SHOR, G.G, JR. 1964. *Marine geology of the Gulf of California*. American Association of Petroleum Geologists, Tulsa, Oklahoma.

VAN WAGONER, J.C., POSAMENTIER, H.W., MITCHUM, R.M., VAIL, P.R., SA R G,J.F., LOUTIT, T.S. & HARDENBOL, J. 1988. An overview of sequence stratigraphy and key definitions. *In*: WILGUS, C.W., HASTINGS, B.S., KENDALL, C.G.ST., POSAMENTIER, H.W., ROSS, C.A. & VAN WAGONER, J.C. (eds), *Sea Level Changes: An Integrated Approach*, 39-45. Society of EconomicPaleontologists and Mineralogists Special Publication, **42**. Tulsa, Oklahoma: Society of Economic Paleontologists and Mineralogists.

VETCH, A.C. & SMITH, P.A. 1939. Atlantic submarine valleys of the United States and the Congo submarine valley. *Geological Society of America Special Paper*, **7**.

VICENTE BRAVO, J.C. & ROBLES, J.S. 1995. Large-scale mesotopographic bedforms from the Albian Black Flysch: characterization, setting and comparison with ancient analogues. *I n*: PICKERING, K.T., HISCOTT, R.N., KENYON, N.H., RICCI LUCCHI, F. & SMITH, R.D.A. (eds), *Atlas of Deep Water Environments: Architectural Style in Turbidite Systems*, 216-226. London: Chapman & Hall.

VINING, B.A., IOANNIDES, N.S. & PICKERING, K.T. 1993. Stratigraphic relationships of some Tertiary lowstand depositional systems in the Central North Sea. *In*: PARKER, J.R. (ed.) *In*: PARKER, J.R. (ed.), *Petroleum Geology of Northwest Europe: Proceedings of the 4th Conference*, 17-30. Bath: The Geological Society.

WALKER, R.G. 1966a. Deep channels in turbidite-bearing formations. *American Association of Petroleum Geologists Bulletin*, **50**, 1899-1917.

WALKER, R.G. 1966b. Shale Grit and Grindslow Shales: transition from turbidite to shallow water sediments in the Upper Carboniferous of Northern England. *Journal of Sedimentary Geology*, **36**, 90-114.

WALKER, R.G. 1975a. Generalized facies model for resedimented conglomerates of turbidite association. *Bulletin of the Geological Society of America*, **86**, 737-748.

WALKER, R.G. 1975b. Upper Cretaceous resedimented conglomerates at Wheeler Gorge, California: description and field guide. *Journal of Sedimentary Petrology*, **45**, 105-112.

WALKER, R.G. 1975c. Nested submarine-fan channels in the Capistrano Formation, San Clemente, California. *Geology Society of America Bulletin*, **86**, 915-924.

WALKER, R.G. 1978. Deep water sandstone facies and ancient submarine fans: models for exploration for stratigraphic traps. *Bulletin of The American Association of Petroleum Geologists*, **62**, 932-966.

WALKER, R.G. 1985. Cardium Formation at Ricinus Field, Alberta: a channel cut and filled by turbiditic currents in Cretaceous western interior seaway. *American Association of Petroleum Geologists*, **69**, 1963-1981.

WALKER, J.R. & MASSINGILL, J.V. 1970. Slump features on the Mississippi fan, northeastern Gulf of Mexico. *Geological Society of America Bulletin*, **81**, 3101-3108.

WALKER, R.G. & E. MUTTI 1973. Turbidite facies and facies associations. *In*: MIDDLETON, G.V. & BOUMA, A.H. (eds), *Turbidites and deep-water sedimentation*, 119-157. Pacific Section Society of Economic Paleontologists and Mineralogists, Short course notes. Tulsa, Oklahoma: Society of Economic Paleontologists and Mineralogists.

WATSON, M.P. 1981. *Submarine fan deposits of the Upper Ordovician-Lower Silurian Milliners Arm Formation, New World Island, Newfoundland.* Unpublished D.Phil. Thesis, University of Oxford, UK.

WEIMER, P. 1991. Sesmic facies, characteristics and variations in channel evolution, Mississippi fan (Plio-Pleistocene), Gulf of Mexico. *In*: WEIMER, P. & LINK, M.H. (eds), *Seismic facies and sedimentary processes of submarine fans and turbidite systems*, 323-348 . New York: Springer-Verlag.

WEIMER, P. & BUFFLER, R.T. 1988. Distribution and seismic facies of Mississippi fan channels. *Geology*, **16**, 900-903.

WETZEL, A. 1993. The transfer of river load to deep-sea fans: a quantitative approach. *American Association of Petroleum Geologists Bulletin*, **77**, 1679-1692.

WILDE, P., NORMARK, W.R. & CHASE, T.E. 1978. Channel sands and petroleum potential of Monterey deep-sea fan, California. *American Association of Petroleum Geologists*, **62**, 967-983.

ZELT, F.B. & ROSSEN, C. 1995. Geometry and continuity of deep-water sandstones and siltstones, Brushy Canyon Formation (Permian) Delaware Mountains, Texas. *In*: PICKERING, K.T., HISCOTT, R.N., KENYON, N.H., RICCI LUCCHI, F. & SMITH, R.D.A. (eds), *Atlas of Deep Water Environments: Architectural Style in Turbidite Systems*, 167-183. London: Chapman & Hall.

Index